Materials, Design and Process Development for Additive Manufacturing

Materials, Design and Process Development for Additive Manufacturing

Editor

Vadim Sufiiarov

MDPI • Basel • Beijing • Wuhan • Barcelona • Belgrade • Manchester • Tokyo • Cluj • Tianjin

Editor
Vadim Sufiiarov
Peter the Great St.Petersburg
Polytechnic University
St. Petersburg
Russia

Editorial Office
MDPI
St. Alban-Anlage 66
4052 Basel, Switzerland

This is a reprint of articles from the Special Issue published online in the open access journal *Materials* (ISSN 1996-1944) (available at: https://www.mdpi.com/journal/materials/special_issues/mater_additive_manufacturing).

For citation purposes, cite each article independently as indicated on the article page online and as indicated below:

LastName, A.A.; LastName, B.B.; LastName, C.C. Article Title. *Journal Name* **Year**, *Volume Number*, Page Range.

ISBN 978-3-0365-4927-9 (Hbk)
ISBN 978-3-0365-4928-6 (PDF)

Cover image courtesy of Vadim Sufiiarov

© 2022 by the authors. Articles in this book are Open Access and distributed under the Creative Commons Attribution (CC BY) license, which allows users to download, copy and build upon published articles, as long as the author and publisher are properly credited, which ensures maximum dissemination and a wider impact of our publications.
The book as a whole is distributed by MDPI under the terms and conditions of the Creative Commons license CC BY-NC-ND.

Contents

About the Editor vii

Vadim Sufiiarov
Special Issue: Materials, Design and Process Development for Additive Manufacturing
Reprinted from: *Materials* **2022**, *15*, 3492, doi:10.3390/ma15103492 1

Xin Ren, Hui Peng, Jingli Li, Hailin Liu, Liming Huang and Xin Yi
Selective Electron Beam Melting (SEBM) of Pure Tungsten: Metallurgical Defects, Microstructure, Texture and Mechanical Properties
Reprinted from: *Materials* **2022**, *15*, 1172, doi:10.3390/ma15031172 5

Gleb A. Turichin, Ekaterina A. Valdaytseva, Stanislav L. Stankevich and Ilya N. Udin
Computer Simulation of Hydrodynamic and Thermal Processes in DLD Technology
Reprinted from: *Materials* **2021**, *14*, 4141, doi:10.3390/ma14154141 15

Marina Gushchina, Gleb Turichin, Olga Klimova-Korsmik, Konstantin Babkin and Lyubov Maggeramova
Features of Heat Treatment the Ti-6Al-4V GTD Blades Manufactured by DLD Additive Technology
Reprinted from: *Materials* **2021**, *14*, 4159, doi:10.3390/ma14154159 25

Ruslan Mendagaliev, Olga Klimova-Korsmik, Vladimir Promakhov, Nikita Schulz, Alexander Zhukov, Viktor Klimenko and Andrey Olisov
Heat Treatment of Corrosion Resistant Steel for Water Propellers Fabricated by Direct Laser Deposition
Reprinted from: *Materials* **2020**, *13*, 2738, doi:10.3390/ma13122738 37

Sergei Ivanov, Marina Gushchina, Antoni Artinov, Maxim Khomutov and Evgenii Zemlyakov
Effect of Elevated Temperatures on the Mechanical Properties of a Direct Laser Deposited Ti-6Al-4V
Reprinted from: *Materials* **2021**, *14*, 6432, doi:10.3390/ma14216432 47

Andrey Samokhin, Nikolay Alekseev, Aleksey Astashov, Aleksey Dorofeev, Andrey Fadeev, Mikhail Sinayskiy and Yulian Kalashnikov
Preparation of W-C-Co Composite Micropowder with Spherical Shaped Particles Using Plasma Technologies
Reprinted from: *Materials* **2021**, *14*, 4258, doi:10.3390/ma14154258 65

Igor Polozov, Nikolay Razumov, Dmitriy Masaylo, Alexey Silin, Yuliya Lebedeva and Anatoly Popovich
Fabrication of Silicon Carbide Fiber-Reinforced Silicon Carbide Matrix Composites Using Binder Jetting Additive Manufacturing from Irregularly-Shaped and Spherical Powders
Reprinted from: *Materials* **2020**, *13*, 1766, doi:10.3390/ma13071766 79

Igor Polozov, Nikolay Razumov, Dmitriy Masaylo, Alexey Silin, Yuliya Lebedeva and Anatoly Popovich
Erratum: Polozov, I., et al. Fabrication of Silicon Carbide Fiber-Reinforced Silicon Carbide Matrix Composites Using Binder Jetting Additive Manufacturing from Irregularly-Shaped and Spherical Powders. *Materials* 2020, *13*, 1766
Reprinted from: *Materials* **2020**, *13*, 2630, doi:10.3390/ma13112630 93

Igor Polozov, Artem Kantyukov, Ivan Goncharov, Nikolay Razumov, Alexey Silin, Vera Popovich, Jia-Ning Zhu and Anatoly Popovich
Additive Manufacturing of Ti-48Al-2Cr-2Nb Alloy Using Gas Atomized and Mechanically Alloyed Plasma Spheroidized Powders
Reprinted from: *Materials* **2020**, *13*, 3952, doi:10.3390/ma13183952 . . . . **95**

Igor Polozov and Anatoly Popovich
Microstructure and Mechanical Properties of NiTi-Based Eutectic Shape Memory Alloy Produced via Selective Laser Melting In-Situ Alloying by Nb
Reprinted from: *Materials* **2021**, *14*, 2696, doi:10.3390/ma14102696 . . . . **111**

Evgenii Borisov, Igor Polozov, Kirill Starikov, Anatoly Popovich and Vadim Sufiiarov
Structure and Properties of Ti/Ti64 Graded Material Manufactured by Laser Powder Bed Fusion
Reprinted from: *Materials* **2021**, *14*, 6140, doi:10.3390/ma14206140 . . . . **123**

Vadim Sufiiarov, Artem Kantyukov, Anatoliy Popovich and Anton Sotov
Structure and Properties of Barium Titanate Lead-Free Piezoceramic Manufactured by Binder Jetting Process
Reprinted from: *Materials* **2021**, *14*, 4419, doi:10.3390/ma14164419 . . . . **133**

Vadim Sufiiarov, Danil Erutin, Artem Kantyukov, Evgenii Borisov, Anatoly Popovich and Denis Nazarov
Structure, Mechanical and Magnetic Properties of Selective Laser Melted Fe-Si-B Alloy
Reprinted from: *Materials* **2022**, *15*, 4121, doi:10.3390/ma15124121 . . . . **147**

About the Editor

Vadim Sufiiarov

Vadim Sufiiarov, Ph.D., is a leading researcher and associate professor at Peter the Great St.Petersburg Polytechnic University. He is majoring in research and development of additive manufacturing (powder bed fusion, directed energy deposition, binder jetting, etc.), new materials, processes and technologies. He obtained his Bachelor's degree in Metallurgy (2006), Master's degree in Materials Science and Technology (2008) and Ph.D. in Material Science and Heat Treatment (2013) from Peter the Great St.Petersburg Polytechnic University. With more than a decade of experience in academia and industry, he has published about 100 papers in scientific journals, presented more than 30 papers at conferences and authored more than 10 patents in areas of metal powder manufacturing, new methods of additive manufacturing and software. Later, he developed teaching disciplines "Introduction to Additive Manufacturing" and "Essence and Features of Additive Manufacturing" for master's students, and he began lecturing in 2015.

Editorial

Special Issue: Materials, Design and Process Development for Additive Manufacturing

Vadim Sufiiarov

Institute of Mechanical Engineering, Materials, and Transport, Peter the Great St. Petersburg Polytechnic University, Polytechnicheskaya 29, 195251 St. Petersburg, Russia; vadim.spbstu@yandex.ru

Additive manufacturing is a dynamically developed direction of modern digital manufacturing processes, which in some cases is already being used to create high-tech products, and in others there are active investigation on new materials and the design and development of technological processes.

Many additive manufacturing processes are based on the use of highly concentrated energy sources, such as laser, electron beam, microplasma, etc., and imply direct manufacture of parts. The most popular processes use powder materials as raw materials, although wire technologies are also gaining popularity at the moment. Local exposure of high-density energy makes it possible to process materials with a high melting point, such as tungsten [1], and to create complex-shaped products from them that were not previously available for manufacture.

Computer modeling expands the understanding of processes occurring during manufacturing and provides information for further improvement of technology. In the article [2], a model has been proposed for the estimation of parameters of a melt pool in laser metal deposition technology (powder-based directed energy deposition process), and the influence of the residual temperature from the previous deposited layers as well as the temperature of the heated powder in the gas–powder jet, taking into account its spatial distribution are considered. Modeling of the size and shape of the melt pool, as well as the profile of its surface, was performed for various alloys: 316 L stainless steel, Inconel 718 nickel alloy and Ti-6Al-4V titanium alloy.

Heat treatment should be applied for most alloys after additive manufacturing to obtain the desired properties. In the articles [3–5], the features of heat treatment on the properties of various application alloys produced by laser metal powder deposition technology are explored. The best mechanical characteristics of additively manufactured Ti-6Al-4V are provided with heat treatment at 900 °C temperatures with a 2-h holding time. The chosen heat treatment mode ensures uniformity of the properties and microstructure in manufactured Ti-6Al-4V samples with variable thickness and complexity, especially in the deposited gas turbine blades [3]. Repeating the cycle of heat treatment twice at T = 620 °C, with a holding time of 2 h for 06Cr15Ni4CuMo steel manufactured by laser powder metal deposition, results in forming a finely dispersed structure of hardened resistant martensite, providing a set of properties equal to the casting material, which makes it promising to use this additive manufacturing process for producing water propellers [4].

The most common technology of powder production for powder-based additive manufacturing processes is gas atomization [6]. At the same time, alternative methods are popular among researchers that allow us to perform synthesis and property evaluation of new materials, with their subsequent application in additive manufacturing processes. The article [7] presents the results of W-C-Co system powder synthesis: In the first stage, agglomerates are formed from nanoparticles by spray-drying technology; then, spherical dense particles are formed by processing in plasma jet. Controlling the plasma treatment process allows us to adjust the technological parameters (average particle size, apparent density, flow rate, etc.) of the synthesized powder material. A similar approach was used

Citation: Sufiiarov, V. Special Issue: Materials, Design and Process Development for Additive Manufacturing. *Materials* **2022**, *15*, 3492. https://doi.org/10.3390/ma15103492

Received: 28 April 2022
Accepted: 11 May 2022
Published: 12 May 2022

Publisher's Note: MDPI stays neutral with regard to jurisdictional claims in published maps and institutional affiliations.

Copyright: © 2022 by the author. Licensee MDPI, Basel, Switzerland. This article is an open access article distributed under the terms and conditions of the Creative Commons Attribution (CC BY) license (https://creativecommons.org/licenses/by/4.0/).

by the authors of the article [8] for the synthesis of SiC-SiC system spherical powder for use in the binder jetting additive manufacturing process. The initial non-spherical powders of SiC and Si were used for synthesis, and after spray drying, non-dense agglomerates of 10–80 µm were formed, bonded with a binding agent (PVA). The process of plasma spheroidization led to the melting of Si, which significantly densified particles, and to the decomposition of PVA binder with the formation of carbon. Carbon reacted with silicon to form a secondary SiC. After plasma treatment, the particles had a spherical shape and were 24–90 µm in size. For the binder jetting, the synthesized powder was mixed with SiC fibers. The results of applying the Ti-48Al-2Cr-2Nb intermetallic alloy powders produced by gas atomization and the combination of mechanical alloying and plasma spheroidization technologies are presented in a previous article [9]. The investigation was carried out via selective laser melting technology (laser powder bed fusion process), and the used SLM system made it possible to implement the process with high-temperature preheating of the substrate (600–900 °C). Crack-free samples were fabricated with a 900 °C preheating temperature. Very fine microstructures consisting of lamellar α_2/γ colonies, equiaxed γ grains, and retained β phase were obtained in all samples. Aluminum loss during the selective laser melting process led to a shift in the solidification route and resulted in the formation of the retained β phase. Increased oxygen content in the initial powder led to the formation of small oxides and an increased α_2 volume fraction. The samples fabricated from gas atomized powder demonstrated superior compressive performance compared to the samples from the mechanically alloyed plasma spheroidized powder. Both alloys showed superior compressive properties compared to the conventional TiAl-alloy. The in situ synthesis approach is often used in additive manufacturing research as an economical method for evaluating microalloying on known compositions, and [10] presented a study of adding niobium to nitinol powder. Despite the incomplete melting of niobium particles, it was possible to achieve the formation of a dense material, with subsequent heat treatment at 900 °C with 2 h of dwell allowing increasing martensitic transformation hysteresis as compared to the alloy without Nb addition from 22 to 50 °C, while the A_f temperature increased from −5 to 22 °C.

A trend that is attracting more and more research attention is multimaterial printing with metals, shown in [11], where different geometries of interfaces, the effects of heat treatment and hot isostatic pressing on the microstructure of intermaterial zones, and mechanical properties were studied. The researchers found that when intermaterial zones are located along the sample, the ratio of the cross-section has greater influence on the mechanical properties than their shape and location. When the zones are arranged transversely to the specimen, a failure occurs at the interface and relative elongation is extremely low.

Some additive manufacturing processes include indirect fabrication, and the formation of the final properties during that type of manufacturing requires applying special post-processing methods (sintering, infiltration, reaction sintering, etc.). Such approaches are used for the additive manufacturing of ceramic materials, as well as for metals [12,13] and composites [14]. The article [8] presents the results of an investigation of manufacturing SiC$_f$-SiC ceramic composite materials. For the manufacture, the binder jetting process was used, after which densification by infiltration and pyrolysis were performed. Polycarbosilane preceramic polymer was used for vacuum infiltration, which has good wettability and small pore filling ability, and after infiltration, the samples were heated up to 1000 °C with dwell for 1 h. As a result, polycarbosilane pyrolysis occurred with the formation of silicon carbide and densification of the samples. The infiltration and pyrolysis procedures were carried out several times. The greatest densification was observed after the first few cycles, and then the rate of rising density (decreasing porosity) became lower. In the article [15], the binder jetting process was used for the fabrication of ceramic green models and pressureless sintering for densification. The effect of sintering modes on the relative density and grain sizes of samples made of two types of materials (submicron powder with unimodal particle-size distribution and micron powder with multimodal particle-size distribution) is presented.

Another promising area is the investigation of materials and processes for the additive manufacturing of functional materials. Shape memory materials are now being actively researched for production by additive manufacturing and expanding possibilities of their application by corrections of alloying systems [10]. Particularly relevant in the global trend towards green energy are materials used in electrical machine component production, such as electricity conductors, soft-magnetic alloys, piezoceramics, etc. Additive manufacturing with such materials is a very current and attractive direction of investigation. Research into the structure and properties of lead-free piezoelectric ceramics based on barium titanate produced by the binder jetting process is presented in the article [15], and the results demonstrate that the use of barium titanate allows achieving high piezoelectric properties, and using binder jetting allows the creation of objects with complex geometry, which has great potential in the manufacture of ultrasonic products used in medicine, aviation, the marine industry, sensors for monitoring welded joints, pressure sensors in pipelines, etc. In [16], three strategies for improving the magnetic properties of Fe-based soft magnetic materials manufactured by selective laser melting have been suggested. The first one is to optimize the parameters of the selective laser melting process for alloys whose chemical composition is chosen to achieve the maximum values of electrical resistivity. Using optimized process parameters and heating of the SLM substrate, the authors obtained a crystalline sample of FeSi6.7 alloy with the minimum coercivity of H_c = 16 A/m and hysteresis losses of 0.7 W/kg at 1 T and 50 Hz. The second one is to optimize the geometry of samples by designing different inner hollow structures into samples. This allows us to reduce eddy current losses in a magnetic field by changing the path of the electric current. The third strategy is to form a sample consisting of alternating layers of soft-magnetic material separated by electrically insulating material (multimaterial printing). The strategies may be combined, which allows more flexible manage properties. Active investigations into the additive manufacturing of amorphous and nanocrystalline soft-magnetic alloys is under development at this time. This type of material has a significantly higher level of soft-magnetic properties compared to crystalline analogues. High cooling rates are typical for the selective laser melting process, but to reduce crystallinity and increase magnetic properties sometimes it is necessary to use nonstandard build strategies of laser scanning [17]. The first strategy was double scanning of each layer, the use of which enabled the researchers to achieve a maximum relative density of 96% and saturation magnetization of 1.22 T. The usage of this scanning strategy also increased the amorphous phase content of samples to 47% and reduced coercivity. The second strategy consists of a two-stage powder melting process: pre-laser melting and short-pulse laser treatment for amorphization. The application of this strategy resulted in a sample containing 89.6% amorphous phase and having a relative density of up to 94.1%.

Funding: The research was supported by a grant from the Russian Science Foundation No. 21-73-10008, https://rscf.ru/project/21-73-10008.

Conflicts of Interest: The author declares no conflict of interest.

References

1. Ren, X.; Peng, H.; Li, J.; Liu, H.; Huang, L.; Yi, X. Selective Electron Beam Melting (SEBM) of Pure Tungsten: Metallurgical Defects, Microstructure, Texture and Mechanical Properties. *Materials* **2022**, *15*, 1172. [CrossRef] [PubMed]
2. Turichin, G.A.; Valdaytseva, E.A.; Stankevich, S.L.; Udin, I.N. Computer simulation of hydrodynamic and thermal processes in DLD technology. *Materials* **2021**, *14*, 4141. [CrossRef] [PubMed]
3. Gushchina, M.; Turichin, G.; Klimova-Korsmik, O.; Babkin, K.; Maggeramova, L. Features of heat treatment the ti-6al-4v gtd blades manufactured by dld additive technology. *Materials* **2021**, *14*, 4159. [CrossRef] [PubMed]
4. Mendagaliev, R.; Klimova-Korsmik, O.; Promakhov, V.; Schulz, N.; Zhukov, A.; Klimenko, V.; Olisov, A. Heat treatment of corrosion resistant steel for water propellers fabricated by direct laser deposition. *Materials* **2020**, *13*, 2738. [CrossRef] [PubMed]
5. Ivanov, S.; Gushchina, M.; Artinov, A.; Khomutov, M.; Zemlyakov, E. Effect of elevated temperatures on the mechanical properties of a direct laser deposited Ti-6Al-4V. *Materials* **2021**, *14*, 6432. [CrossRef]
6. Golod, V.M.; Sufiiarov, V.S. The evolution of structural and chemical heterogeneity during rapid solidification at gas atomization. *IOP Conf. Ser. Mater. Sci. Eng.* **2017**, *192*, 12009. [CrossRef]

7. Samokhin, A.; Alekseev, N.; Astashov, A.; Dorofeev, A.; Fadeev, A.; Sinayskiy, M.; Kalashnikov, Y. Preparation of w-c-co composite micropowder with spherical shaped particles using plasma technologies. *Materials* **2021**, *14*, 4258. [CrossRef] [PubMed]
8. Polozov, I.; Razumov, N.; Masaylo, D.; Silin, A.; Lebedeva, Y.; Popovich, A. Fabrication of silicon carbide fiber-reinforced silicon carbide matrix composites using binder jetting additive manufacturing from irregularly-shaped and spherical powders. *Materials* **2020**, *13*, 1766. [CrossRef] [PubMed]
9. Polozov, I.; Kantyukov, A.; Goncharov, I.; Razumov, N.; Silin, A.; Popovich, V.; Zhu, J.N.; Popovich, A. Additive manufacturing of Ti-48Al-2Cr-2Nb alloy using gas atomized and mechanically alloyed plasma spheroidized powders. *Materials* **2020**, *13*, 3952. [CrossRef] [PubMed]
10. Polozov, I.; Popovich, A. Microstructure and mechanical properties of niti-based eutectic shape memory alloy produced via selective laser melting in-situ alloying by nb. *Materials* **2021**, *14*, 2696. [CrossRef] [PubMed]
11. Borisov, E.; Polozov, I.; Starikov, K.; Popovich, A.; Sufiiarov, V. Structure and properties of Ti/Ti64 graded material manufactured by laser powder bed fusion. *Materials* **2021**, *14*, 6140. [CrossRef] [PubMed]
12. Polozov, I.; Sufiiarov, V.; Shamshurin, A. Synthesis of titanium orthorhombic alloy using binder jetting additive manufacturing. *Mater. Lett.* **2019**, *243*, 88–91. [CrossRef]
13. Sufiiarov, V.; Polozov, I.; Kantykov, A.; Khaidorov, A. Binder jetting additive manufacturing of 420 stainless steel: Densification during sintering and effect of heat treatment on microstructure and hardness. *Mater. Today Proc.* **2019**, *30*, 592–595. [CrossRef]
14. Sufiiarov, V.; Kantyukov, A.; Polozov, I. Reaction sintering of metal-ceramic AlSI-Al$_2$O$_3$ composites manufactured by binder jetting additive manufacturing process. In Proceedings of the METAL 2020—29th International Conference on Metallurgy and Materials, Brno, Czech Republic, 20–22 May 2020; pp. 1148–1155.
15. Sufiiarov, V.; Kantyukov, A.; Popovich, A.; Sotov, A. Structure and properties of barium titanate lead-free piezoceramic manufactured by binder jetting process. *Materials* **2021**, *14*, 4419. [CrossRef] [PubMed]
16. Goll, D.; Schuller, D.; Martinek, G.; Kunert, T.; Schurr, J.; Sinz, C.; Schubert, T.; Bernthaler, T.; Riegel, H.; Schneider, G. Additive manufacturing of soft magnetic materials and components. *Addit. Manuf.* **2019**, *27*, 428–439. [CrossRef]
17. Ozden, M.G.; Morley, N.A. Laser additive manufacturing of Fe-based magnetic amorphous alloys. *Magnetochemistry* **2021**, *7*, 20. [CrossRef]

Article

Selective Electron Beam Melting (SEBM) of Pure Tungsten: Metallurgical Defects, Microstructure, Texture and Mechanical Properties

Xin Ren [1], Hui Peng [2], Jingli Li [1], Hailin Liu [1], Liming Huang [1] and Xin Yi [1,*]

[1] Department of Mechanics and Engineering Science, College of Engineering, Peking University, Beijing 100871, China; xinren_1986@163.com (X.R.); lijingli@pku.edu.cn (J.L.); 1601111544@pku.edu.cn (H.L.); liming_huang@pku.edu.cn (L.H.)

[2] Research Institute for Frontier Science, Beihang University, Beijing 100191, China; penghui@buaa.edu.cn

* Correspondence: xyi@pku.edu.cn

Citation: Ren, X.; Peng, H.; Li, J.; Liu, H.; Huang, L.; Yi, X. Selective Electron Beam Melting (SEBM) of Pure Tungsten: Metallurgical Defects, Microstructure, Texture and Mechanical Properties. *Materials* **2022**, *15*, 1172. https://doi.org/10.3390/ma15031172

Academic Editors: Vadim Sufiiarov and Konda Gokuldoss Prashanth

Received: 1 January 2022
Accepted: 31 January 2022
Published: 3 February 2022

Publisher's Note: MDPI stays neutral with regard to jurisdictional claims in published maps and institutional affiliations.

Copyright: © 2022 by the authors. Licensee MDPI, Basel, Switzerland. This article is an open access article distributed under the terms and conditions of the Creative Commons Attribution (CC BY) license (https://creativecommons.org/licenses/by/4.0/).

Abstract: Effects of processing parameters on the metallurgical defects, microstructure, texture, and mechanical properties of pure tungsten samples fabricated by selective electron beam melting are investigated. SEBM-fabricated bulk tungsten samples with features of lack of fusion, sufficient fusion, and over-melting are examined. For samples upon sufficient fusion, an ultimate compressive strength of 1.76 GPa is achieved at the volumetric energy density of 900 J/mm^3–1000 J/mm^3. The excellent compressive strength is higher and the associated volumetric energy density is significantly lower than corresponding reported values in the literature. The average relative density of SEBM-fabricated samples is 98.93%. No microcracks, but only pores with diameters of few tens of micrometers, are found in SEBM-ed tungsten samples of sufficient fusion. Properties of samples by SEBM and selective laser melting (SLM) have also been compared. It is found that SLM-fabricated samples exhibit inevitable microcracks, and have a significantly lower ultimate compressive strength and a slightly lower relative density of 98.51% in comparison with SEBM-ed samples.

Keywords: pure tungsten; selective electron beam melting (SEBM); porosity; microstructure; mechanical properties

1. Introduction

Tungsten has a high melting point, a high density, a low erosion tendency, high thermal stress resistance, and high thermal conductivity, and exhibits low swelling and low tritium retention. Owing to these superior thermophysical properties, tungsten and tungsten alloys are considered the most promising candidate materials for plasma facing components in nuclear reactors [1], and have also been attractive for industrial, aerospace, and medical applications such as high temperature furnaces and X-ray shielding [2]. On the other hand, tungsten materials show processing difficulties and have limited engineering applications due to the inherent features of high-melting point, brittleness at room temperature, and high temperature oxidation behaviors [3,4].

Selective laser melting (SLM) and selective electron beam manufacturing (SEBM) are two important representatives of powder bed based additive manufacturing processes for forming near-net shaped components of metals. They not only offer new feasible processing methods for refractory metals, but also liberate the design freedom and significantly expand the engineering application scope of refractory metals.

Rapid progress has been made in understanding the processing-microstructure-properties relationships of pure tungsten samples fabricated by SLM [3–7]. For example, the SLM-fabricated bulk pure tungsten samples of high relative densities of 95–98.51% can achieve an ultimate compressive strength up to 1.01 GPa after heat treatment [7,8]. Experimental observations also indicate that microcracks owing to high residual stresses induced by high temperature gradients in SLM-ed samples are inevitable [4–11]. In SLM-ed samples, one can also find pores whose formation is most likely attributed to the entrapped protective inert gas (e.g., argon) by the Marangoni effect [7,12].

In contrast to SLM with laser power on the order of hundreds of Watts, the electron beam power of SEBM can reach as high as 3 kW, much higher energy input than SLM [13]. In SEBM, preheating the substrate and every powder layer by the defocused electron beam can significantly minimize the residual stress in the fabricated components. The SEBM process works under a vacuum condition which forms a nearly perfect protection against oxidization and gas contamination [14,15]. Owing to these features, SEBM is more suitable than SLM for fabricating refractory metal components and brittle materials with affinity to gases such as oxygen at high temperature [16–19].

In comparison with studies of SLM-ed pure tungsten samples [3–7], few studies on the fabrication of pure tungsten by SEBM have been reported in the literature [20]. A processing window for SEBM-fabricated pure tungsten is preliminarily determined by characterizing the surface morphologies of the melt pool, and the sample compression strength along the building direction is reported as 1.56 GPa [20]. Nevertheless, important questions, such as how to relate the mechanical properties of SEBM-ed pure tungsten samples to the microstructure and texture, remain to be fully elucidated. Moreover, no comparative analysis has been performed on the microstructural and mechanical characterizations of pure tungsten samples manufactured by SEBM or SLM.

In this work, we focus on some key issues of SEBM-ed pure tungsten samples, such as features of microdefects including cracks and pores, microstructure, texture, and mechanical properties. A thorough comparison on the microstructure and mechanical properties of pure tungsten samples manufactured by SEBM or SLM is performed, which is helpful in understanding and optimizing the additive manufacture of tungsten samples.

2. Materials and Experimental Procedures

2.1. Materials and SEBM Process

Spherical powder particles of good flowability are most suitable for SEBM. However, owing to a large amount of powder required for SEBM and the corresponding expensive cost attributed to spheroidization of pure tungsten powder, polygonal pure tungsten powder of good flowability emerge as an alternative option. The Hall flow rate of polygonal pure tungsten powder particles with size of 65 μm–105 μm (Figure 1) is measured according to the ASTM B213 standard test method. Measurements with 50 g mass of powder require about 8.5 s, indicating fluidity good enough for SEBM.

Figure 1. Morphology of tungsten powders used in the SEBM process.

The Arcam A2XX EBM system (Mölndal, Sweden) is used to fabricate the bulk pure tungsten samples with a minimum electron beam diameter of around 250 μm. The vacuum pressure in the chamber is kept below 0.2 Pa. The 316 stainless steel substrate plate is initially preheated to 1150 °C by fast scanning with a defocused electron beam to decrease thermal gradients during the building process. Then, a sample is fabricated layer by layer, and each layer is subject to a four-step process of depositing-preheating-melting a powder layer and lowering platform (Figure 2a): (1) depositing a powder layer onto the substrate plate by the rake, (2) preheating and slightly sintering the powder layer with a defocused electron beam, (3) selectively scanning and melting the preheated powder layer by a focused electron beam according to a schemed computer-aided system, and (4) lowering the building platform by a nominal layer thickness and repeating from step (1) for the next layer fabrication. Step (2) is essential to prevent so-called smoke events occurring in step (3), which lead to unfavorable powder spreading within the machine and eventual termination of the SEBM process [21]. A zigzag scanning pattern with an interlayer misorientation of 90° between layers is adopted (Figure 2b).

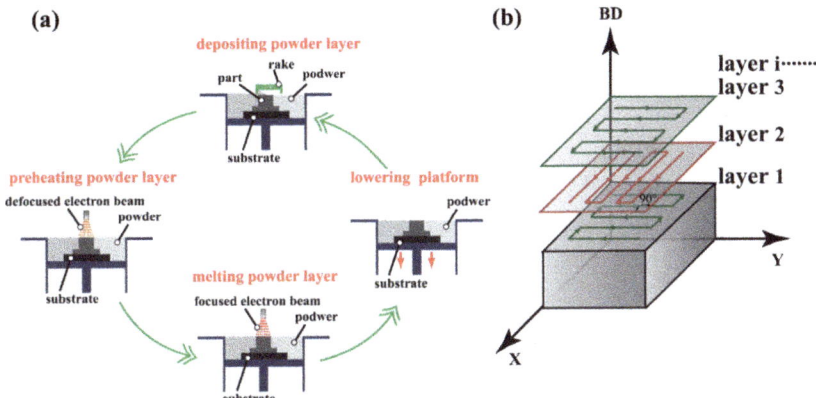

Figure 2. Schematic of the four-step process for building each layer in SEBM (**a**) and zigzag scanning strategy (**b**). BD, building direction.

Critical processing parameters impacting the input energy density in SEBM include the electron beam power P, scanning velocity v, powder layer thickness t, and hatch distance h. A combination of them gives the volumetric energy density $E = P/(h \times v \times t)$, which is used to estimate the energy density input into the powder layer. The beam power is given by $P = I \times U$, where I is the beam current and U is the voltage. The building processes exhibit features of lack of fusion, sufficient fusion, and over-melting as E increases. A lower E with a high scanning speed may induce the incomplete spread of the liquid metal and lead to the lack of fusion. An excessive E with a low scanning speed could result in the disturbance of the melt pool. Values of the processing parameters are listed in Table 1. The SEBM-ed pure tungsten samples have dimension of $15 \times 15 \times 20$ mm^3 (Figure 3).

Table 1. Processing parameters for the fabrication of tungsten using SEBM.

	Beam Current I	Layer Thickness t	Hatch Distance h	Voltage U
	15 mA	50 µm	100 µm	60 kV
No. [1]	Scanning Velocity v (mm/s)		Volumetric Energy Density E (J/mm^3)	
S1	300		600	
S2	250		720	
S3	200		900	
S4	180		1000	
S5	150		1200	

[1] S1 and S2, lack of fusion; S3 and S4, sufficient fusion; S5, over-melting.

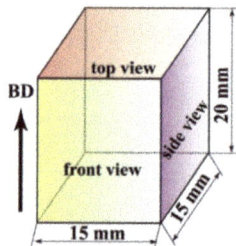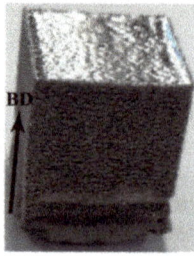

Figure 3. Pure tungsten cube (S4) of size 15 × 15 × 20 mm³ fabricated by SEBM.

2.2. Characterization

The SEBM-ed bulk pure tungsten samples are cut from the substrate using wire electrical discharge machining, then cleaned with acetone, alcohol, and water. For microstructure characterization, smaller specimens cut from the bulk components are mechanically ground, followed by electrolytic polishing with 2% NaOH solution at voltage of 20 V, then these specimens are examined with electron backscatter diffraction (EBSD) analysis. To detect the metallographic microstructure, the samples undergo electrolytic etching with 2% NaOH solution at voltage of 5 V and are examined using the Leica DM2700 M optical microscope (Wetzlar, Germany). Cylinder samples with length-to-diameter ratio $L/D = 1.5$ are cut from the fabricated components and the end faces are polished for compression tests at room temperature using an Instron 5585H universal testing equipment (Norwood, United States) at a strain rate of 10^{-3}/s.

3. Results and Discussion

3.1. Metallurgical Defects

Three representative top surface morphologies and side view optical micrographs of the bulk pure tungsten samples fabricated by SEBM are shown in Figure 4. At high scanning velocity and low volumetric energy density, the liquid metal has no sufficient time to fully spread owing to the rapid solidification velocity, and evident shrinkage holes could be formed. As shown in Figure 4a, the corresponding fabricated samples show features of lack of fusion with large shrinkage holes having rough surfaces. As the scanning velocity v decreases to a proper value, samples of smooth surfaces and dense structures with few tiny pores are fabricated with sufficient fusion (Figure 4b). As v further decreases, the energy input becomes excessive, causing over-melting and unfavorable disturbance of the melt pool and leading to a higher surface periphery (Figure 4c). Although the energy inputs for the samples of sufficient fusion and over-melting are different, the pores there have similar size and number. The pore formation might be due to the trapping of gas such as residual oxygen within the powder particles, even the building process works in a high vacuum environment. A thorough and detailed mechanistic study on the formation of different surface morphologies is challenging and deserves further detailed investigations in the future. The effects of pores on the mechanical properties of the SEBM-ed samples are discussed comparing with SLM-ed samples in Section 3.3.

No microcracks are seen; only pores (irregular shrinkage holes in the case of lack of fusion, and tiny spherical pores in cases of sufficient fusion and over-melting) exist in the SEBM-ed tungsten samples (Figure 4). In SLM-ed tungsten samples, microcracks are usually inevitable [6,8–10], as shown in Figure 5. A widely adopted aspect causing the microcrack formation in the SLM-ed samples is the high thermal stress generated by the high temperature gradient during the SLM process. That high thermal stress could cause tearing along grain boundaries, which are embrittled by the oxidation and resulting impurities [4,9,22,23]. It has been reported that a high concentration of oxygen and impurities segregate to the grain boundaries during the cooling process of melted tungsten in SLM [9,24], and the strength of pure tungsten and tungsten alloys are greatly affected by the oxides distributed at grain boundaries [22,25–28].

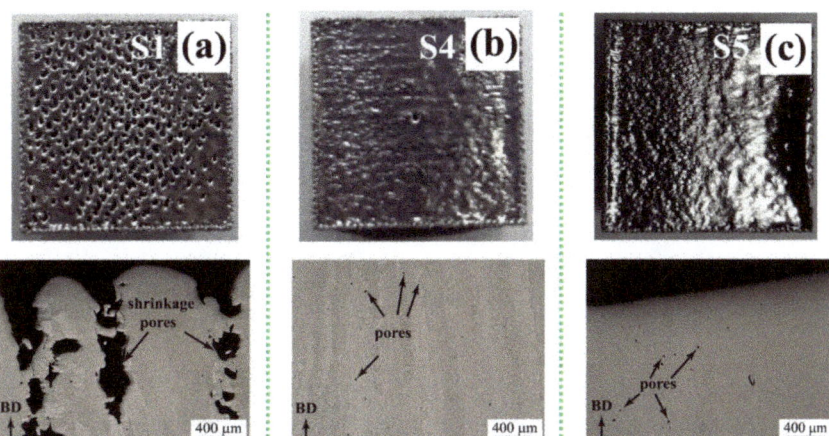

Figure 4. Three representative top surface morphologies (top row) and side view optical micrographs (bottom row) of the bulk pure tungsten samples (S1, S4, and S5) fabricated by SEBM with characteristics of lack of fusion (**a**), sufficient fusion (**b**), and over-melting or excess energy input (**c**). In (**c**), the surface periphery is higher than the central region. Lateral dimension is 15 × 15 mm^2.

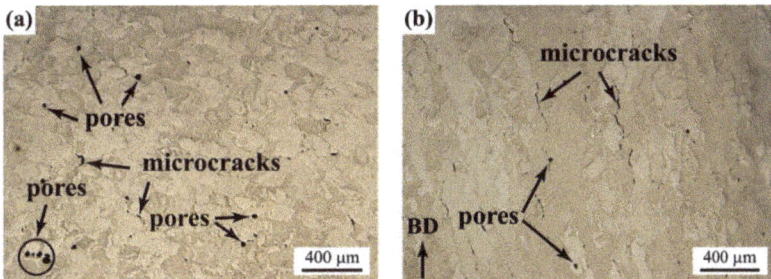

Figure 5. Optical micrographs of polished cross-sections (top (**a**) and front (**b**) views) for bulk pure tungsten samples fabricated by SLM.

In SEBM, the vacuum environment prohibits the oxidation, and the high temperature gradient is lower compared to SLM. Therefore, no microcracks are formed in SEBM-ed samples. For example, in SEBM for tungsten fabrication the substrate is heated to 1150 °C by the defocused electron beam and that temperature is maintained throughout the building process. Each powder layer is also preheated. Therefore, the temperature gradient in SEBM is much lower than that in SLM for tungsten where the substrate can only be pre-heated up to 200 °C and there is no preheating for powder layers [8]. The lower temperature gradient in SEBM gives rise to a lower cooling rate, which in turn causes a much lower thermal stress [13,29,30] and low possibility of microcrack formation. Moreover, the ductile-to-brittle transition temperature (DBTT) of pure tungsten is in a range of 150 °C to 400 °C [1,31]. The sustained elevated temperature in SEBM could not only relieve the thermal stress [32], but also enable SEBM-ed tungsten samples above DBTT exhibiting plasticity to a certain extent. Overall, the crack formation in SEBM-ed tungsten is inhibited during the building process.

According to the Archimedes' principle, relative densities of as-fabricated SEBM samples S1–S5 are measured. Having knowledge of the volumetric energy density E in

Table 1, the relation between the relative density and the volumetric energy density is determined (Figure 6). It is shown that samples of either sufficient fusion or over-melting have high relative densities, and an S4 sample of sufficient fusion has the highest relative density of 98.93%. Though S5 samples have high relative densities, the associated irregular surface periphery, as shown in Figure 4c, limits the usage of S5 samples of over-melting as near-net shaped components.

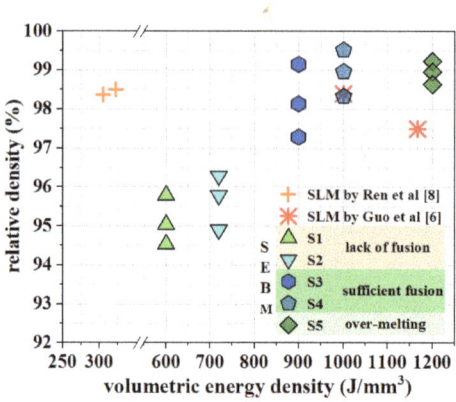

Figure 6. The relative density versus the volumetric energy density of the pure tungsten additive fabrication.

In Figure 6, we also provide the values associated with SLM-ed tungsten for references. In our previous work on the additive manufacture of pure tungsten samples by SLM with spherical powder particles of 5 μm to 25 μm in diameter, the optimized SLM processing parameters include layer thickness of 30 μm, hatch distance of 0.08 mm, laser power of 300 W and scanning velocity of 300 mm/s at a laser spot size of around 70 μm [8]. As in the SLM-ed samples there only exist small microcracks and the crack number is small (Figure 5), the relative density of the SLM-ed samples is only slightly lower than the SEBM-ed samples of sufficient fusion with the absence of microcracks. In SLM and SEBM, the energy absorption depends on the powder particle size and shape as well as the physical features of laser and electron beams [33,34]. As all these parameters in the present work of SEBM and our previous work of SLM are significantly different, it is not surprising that the volumetric energy densities of the SLM-ed and SEBM-ed tungsten samples with similar relative densities could be significantly different (Figure 6).

3.2. Microstructure and Texture

Columnar grains are observed in all SEBM-ed samples (Figure 7). It is known that the columnar grains grow along the maximum temperature gradient direction. In most cases, the building direction is parallel to the direction of the maximum temperature gradient, and one can see that columnar grains grow epitaxially along the building direction for SEBM-ed samples (as demonstrated in Figure 7a,c). In some cases, particularly at the sample edges surrounded by thick powder layers, the heat dissipation rate of the particles is much lower than that of the bulk, resulting in an evident deviation of the maximum temperature gradient direction from the building direction. That direction deviation is gradually reduced toward the bulk region, consistent with marked arrows in Figure 7b.

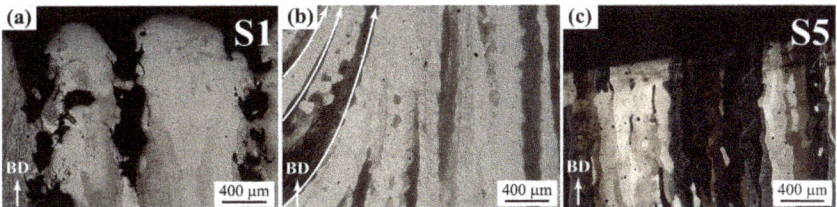

Figure 7. Columnar grain structures of as-fabricated pure tungsten samples of lack of fusion (**a**), sufficient fusion (**b**), and over-melting (**c**).

Microstructure and crystallographic texture analysis of the SEBM-ed sample S4 is performed using EBSD analysis. As shown in Figure 8, coarse grains appear equiaxed from the top view (Figure 8a), significantly different from the scattered checkboard pattern of SLM-ed samples reported in our previous work [8]. Long columnar grain structures along the building direction are observed in SEBM-ed samples (Figure 8a), indicating a stable melt pool during the SEBM process [20]. In contrast, columnar grain structures in the SLM-ed samples are relatively small (grain diameter about 125 μm in comparison with 320 μm for the SEBM-ed samples) and discrete [8], indicating disrupted epitaxy growth of columnar grains during SLM, which might be owing to the scanning strategy of a pattern with an interlayer misorientation of around 67° [8,10].

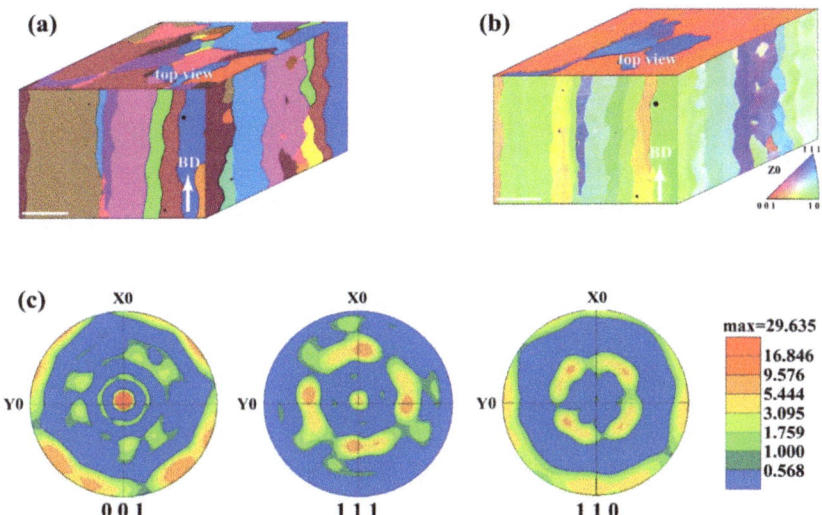

Figure 8. EBSD analysis of the SEBM-ed sample S4. (**a**) Grain size characterization, (**b**) inverse pole figure (IPF) coloring map, and (**c**) orientation pole figure taken from top view in (**b**). Scale bars in (**a**,**b**), 200 μm.

The IPF coloring map in Figure 8b and the pole figure in Figure 8c show that the SEBM-ed pure tungsten S4 sample has a strong <100> grain orientations. As pure tungsten has a body-centered cubic crystal structure, the columnar grains usually prefer <100> growth direction [35], consistent with Figure 8c. In the process of melting and solidification of tungsten, the epitaxial growth of columnar grains shows a strong rotated cube {100} <110> along the building direction. An unusual (111) texture has been found in SLM-ed tungsten, which is owing to the rotation of thermal gradient caused by the 67° rotation of scanning direction between layers [29,36]. The scanning direction with rotation of 90° between layers adopted in the present SEBM process may have little influence on the direction of maximum thermal gradient, and therefore the preferred <100> growth direction is observed in this work.

3.3. Mechanical Properties

Pore formation affects not only the density but also the mechanical properties of fabricated samples. Compression tests of cylindrical tungsten samples with features of lack of fusion (S1), sufficient fusion (S4), and over-melting (S5) are conducted at room temperature (Figure 9). The cylindrical samples have diameter of 3 mm and length of 4.5 mm. The average ultimate compressive strength for each sample is given in the inset of Figure 9.

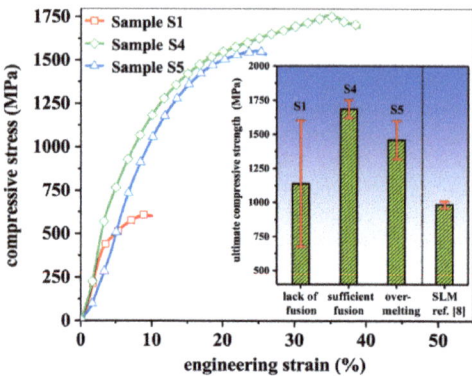

Figure 9. Compressive stress–strain curves for SEBM-ed pure tungsten samples S1, S4, and S5. Inset, ultimate compressive strengths of corresponding pure tungsten SEBM-ed bulk samples and previous SLM-ed sample.

Owing to the large and irregular pore structures in S1 samples (Figures 4a and 7a), the corresponding ultimate compressive strength is as small as 0.61 GPa and the fracture engineering strain is as small as around 10%. Samples S4 and S5 of dense structures containing tiny pores (Figures 4b,c and 7b,c) have a much larger ultimate compressive strength and fracture strain. Specifically, the ultimate compressive strength of the S4 sample (sufficient fusion) is 1.76 GPa with a fracture engineering strain of around 40%, both larger than reported values of 1.56 GPa and 18% in the literature [20]. Moreover, the volumetric energy density for the S4 sample is about 1000 J/mm^3, significantly lower than that of 1440 J/mm^3–3840 J/mm^3 in ref. [20]. The compressive strength of optimized SLM-ed bulk pure tungsten samples is around 1.01 GPa [8], much lower than that achieved in the present work by SEBM, though the SLM-ed sample also has a high relative density of 98.51%. This observation suggests that the large ultimate compressive strength of SEBM-ed S4 sample mainly results from the absence of microcracks in fabricated samples, rather than the high relative density. In S4, samples the columnar grains are relatively coarse due to the long-time high temperature fabrication process, and grain refinement with addition of the carbide nanoparticles or oxide dispersion [9,22,23,26,37] could be employed to further improve the mechanical properties of fabricated samples.

4. Conclusions

Pure tungsten samples of high relative density and superior ultimate compressive strength are fabricated by SEBM at different volumetric energy densities. Samples with features of lack of fusion, sufficient fusion, and over-melting are identified. The corresponding metallurgical defects, microstructure, texture, and mechanical properties of the SEBM-fabricated samples are analyzed and compared with counterparts of the SLM-fabricated samples. The main conclusions are summarized as follows.

1. Pure tungsten bulk samples upon sufficient fusion using SEBM are fabricated with high relative density up to 98.93%. A superior ultimate compressive strength of 1.76 GPa higher than reported values in literature is achieved at a volumetric energy density of 1000 J/mm^3.
2. Long columnar grain structures along the building direction are observed in the SEBM-ed samples, indicating a stable melt pool during the SEBM process; while relatively small and discrete columnar grain structures exist in the SLM-ed samples, indicating disrupted epitaxy growth of columnar grains during SLM.
3. Comparing with the SLM-ed pure tungsten samples containing micropores and inevitable microcracks, no microcracks, but only micropores, are found in the SEBM-ed tungsten samples of sufficient fusion. The absence of microcracks in the SEBM-ed tungsten benefits from the reduction in oxide precipitates due to the vacuum manufacturing environment and from the low thermal stress owing to the specific heating and preheating schemes in the SEBM process.
4. The SEBM-ed samples of sufficient fusion have the ultimate compressive strength (1.76 GPa) significantly larger than that (0.98 GPa) of the SLM-ed samples of similar high relative densities exceeding 98%.

Author Contributions: Conceptualization, X.R. and X.Y.; methodology, X.R. and H.P.; investigation, X.R., H.P., J.L., H.L., L.H. and X.Y.; writing—original draft preparation, X.R.; writing—review and editing, X.Y.; supervision, X.Y.; and funding acquisition, X.Y. All authors have read and agreed to the published version of the manuscript.

Funding: This research was funded by the National Natural Science Foundation of China (grant No. 11988102) and National Science and Technology Major Project (2017-VI-0003-0073).

Conflicts of Interest: The authors declare no conflict of interest.

References

1. Philipps, V. Tungsten as material for plasma-facing components in fusion devices. *J. Nucl. Mater.* **2011**, *415*, S2–S9. [CrossRef]
2. Lassner, E.; Schuber, W.-D. *Tungsten—Properties, Chemistry, Technology of the Element, Alloys, and Chemical Compounds*; Springer: Berlin/Heidelberg, Germany, 1999.
3. Zhang, D.; Cai, Q.; Liu, J. Formation of nanocrystalline tungsten by selective laser melting of tungsten powder. *Mater. Manuf. Process.* **2012**, *27*, 1267–1270. [CrossRef]
4. Iveković, A.; Omidvari, N.; Vrancken, B.; Lietaert, K.; Thijs, L.; Vanmeensel, K.; Vleugels, J.; Kruth, J.-P. Selective laser melting of tungsten and tungsten alloys. *Int. J. Refract. Met. Hard Mater.* **2018**, *72*, 27–32. [CrossRef]
5. Guo, M.; Gu, D.; Xi, L.; Zhang, H.; Zhang, J.; Yang, J.; Wang, R. Selective laser melting additive manufacturing of pure tungsten: Role of volumetric energy density on densification, microstructure and mechanical properties. *Int. J. Refract. Met. Hard Mater.* **2019**, *84*, 105025. [CrossRef]
6. Guo, M.; Gu, D.; Xi, L.; Du, L.; Zhang, H.; Zhang, J. Formation of scanning tracks during Selective Laser Melting (SLM) of pure tungsten powder: Morphology, geometric features and forming mechanisms. *Int. J. Refract. Met. Hard Mater.* **2019**, *79*, 37–46. [CrossRef]
7. Tan, C.; Zhou, K.; Ma, W.; Attard, B.; Zhang, P.; Kuang, T. Selective laser melting of high-performance pure tungsten: Parameter design, densification behavior and mechanical properties. *Sci. Technol. Adv. Mat.* **2018**, *19*, 370–380. [CrossRef]
8. Ren, X.; Liu, H.; Lu, F.; Huang, L.; Yi, X. Effects of processing parameters on the densification, microstructure and mechanical properties of pure tungsten fabricated by optimized selective laser melting: From single and multiple scan tracks to bulk parks. *Int. J. Refract. Met. Hard Mater.* **2021**, *96*, 105490. [CrossRef]
9. Li, K.; Wang, D.; Xing, L.; Wang, Y.; Yu, C.; Chen, J.; Zhang, T.; Ma, J.; Liu, W.; Shen, Z. Crack suppression in additively manufactured tungsten by introducing secondary-phase nanoparticles into the matrix. *Int. J. Refract. Met. Hard Mater.* **2019**, *79*, 158–163. [CrossRef]
10. Wang, D.-Z.; Li, K.-L.; Yu, C.-F.; Ma, J.; Liu, W.; Shen, Z.-J. Cracking behavior in additively manufactured pure tungsten. *Acta Metall. Sin.* **2019**, *32*, 127–135. [CrossRef]
11. Müller, A.v.; Schlick, G.; Neu, R.; Anstätt, C.; Klimkait, T.; Lee, J.; Pascher, B.; Schmitt, M.; Seidel, C. Additive manufacturing of pure tungsten by means of selective laser beam melting with substrate preheating temperatures up to 1000 °C. *Nucl. Mater. Energy* **2019**, *19*, 184–188. [CrossRef]
12. Ng, G.K.L.; Jarfors, A.E.W.; Bi, G.; Zheng, H.Y. Porosity formation and gas bubble retention in laser metal deposition. *Appl. Phys. A* **2009**, *97*, 641–649. [CrossRef]

13. Körner, C. Additive manufacturing of metallic components by selective electron beam melting—A review. *Int. Mater. Rev.* **2016**, *61*, 361–377. [CrossRef]
14. Gong, X.; Anderson, T.; Chou, K. Review on powder-based electron beam additive manufacturing technology. *Manuf. Rev.* **2014**, *1*, 2. [CrossRef]
15. Gokuldoss, P.K.; Kolla, S.; Eckert, J. Additive manufacturing processes: Selective laser melting, electron beam melting and binder jetting—Selection guidelines. *Materials* **2017**, *10*, 672. [CrossRef]
16. Yang, J.; Huang, Y.; Liu, B.; Guo, C.; Sun, J. Precipitation behavior in a Nb-5W-2Mo-1Zr niobium alloy fabricated by electron beam selective melting. *Mater. Charact.* **2021**, *174*, 111019. [CrossRef]
17. Higashi, M.; Yoshimi, K. Electron beam surface melting of MoSiBTiC alloys: Effect of preheating on cracking behavior and microstructure evolution. *Mater. Des.* **2021**, *209*, 110010. [CrossRef]
18. Yue, H.; Peng, H.; Li, R.; Qi, K.; Zhang, L.; Lin, J.; Su, Y. Effect of heat treatment on the microstructure and anisotropy of tensile properties of TiAl alloy produced via selective electron beam melting. *Mater. Sci. Eng. A* **2021**, *803*, 140473. [CrossRef]
19. Fernandez-Zelaia, P.; Ledford, C.; Ellis, E.A.I.; Campbell, Q.; Rossy, A.M.; Leonard, D.N.; Kirka, M.M. Crystallographic texture evolution in electron beam melting additive manufacturing of pure Molybdenum. *Mater. Des.* **2021**, *207*, 109809. [CrossRef]
20. Yang, G.; Yang, P.; Yang, K.; Liu, N.; Jia, L.; Wang, J.; Tang, H. Effect of processing parameters on the density, microstructure and strength of pure tungsten fabricated by selective electron beam melting. *Int. J. Refract. Met. Hard Mater.* **2019**, *84*, 105040. [CrossRef]
21. Zhong, Y.; Rännar, L.-E.; Liu, L.; Koptyug, A.; Wikman, S.; Olsen, J.; Cui, D.; Shen, Z. Additive manufacturing of 316L stainless steel by electron beam melting for nuclear fusion applications. *J. Nucl. Mater.* **2017**, *486*, 234–245. [CrossRef]
22. Wang, R.; Xie, Z.M.; Liu, R.; Gao, R.; Yang, J.F.; Fang, Q.F.; Liu, C.S.; Wu, X.B.; Wang, T.; Song, X.P.; et al. Effects of ZrC content on the mechanical properties and microstructures of hot-rolled W-ZrC composites. *Nucl. Mater. Energy* **2019**, *20*, 100705. [CrossRef]
23. Hu, Z.; Zhao, Y.; Guan, K.; Wang, Z.; Ma, Z. Pure tungsten and oxide dispersion strengthened tungsten manufactured by selective laser melting: Microstructure and cracking mechanism. *Addit. Manuf.* **2020**, *36*, 101579. [CrossRef]
24. Xie, Z.M.; Liu, R.; Miao, S.; Yang, X.D.; Zhang, T.; Wang, X.P.; Liu, X.; Lian, I.; Luo, N.; Fang, F.; et al. Extraordinary high ductility/strength of the interface designed bulk W-ZrC alloy plate at relatively low temperature. *Sci. Rep.* **2015**, *5*, 16014. [CrossRef] [PubMed]
25. Zhang, T.; Wang, Y.K.; Xie, Z.M.; Liu, C.S.; Fang, Q.F. Synergistic effects of trace ZrC/Zr on the mechanical properties and microstructure of tungsten as plasma facing materials. *Nucl. Mater. Energy* **2019**, *19*, 225–229. [CrossRef]
26. Liu, R.; Xie, Z.M.; Yang, J.F.; Zhang, T.; Hao, T.; Wang, X.P.; Fang, Q.F.; Liu, C.S. Recent progress on the R&D of W-ZrC alloys for plasma facing components in fusion devices. *Nucl. Mater. Energy* **2018**, *16*, 191–206.
27. Miao, S.; Xie, Z.; Lin, Y.; Fang, Q.; Tan, J.; Zhao, Y. Mechanical properties, thermal stability and microstructures of W-Re-ZrC alloys fabricated by spark plasma sintering. *Metals* **2020**, *10*, 277. [CrossRef]
28. Gludovatz, B.; Wurster, S.; Weigärtner, T.; Hoffmann, A.; Pippan, R. Influence of impurities on the fracture behaviour of tungsten. *Philos. Mag.* **2011**, *91*, 3006–3020. [CrossRef]
29. Wright, J. Additive Manufacturing of Tungsten via Selective Laser Melting and Electron Beam Melting. Ph.D. Thesis, The University of Sheffield, Sheffield, UK, 2020.
30. Sochalski-Kolbus, L.M.; Payzant, E.A.; Cornwell, P.A.; Watkins, T.R.; Babu, S.S.; Dehoff, R.R.; Lorenz, M.; Ovchinnikova, O.; Duty, C. Comparison of residual stresses in Inconel 718 simple parts made by electron beam melting and direct laser metal sintering. *Metall. Mater. Trans. A* **2015**, *46A*, 1419–1432. [CrossRef]
31. Ren, C.; Fang, Z.Z.; Koopman, M.; Butler, B.; Paramore, J.; Middlemas, S. Methods for improving ductility of tungsten—A review. *Int. J. Refract. Met. Hard Mater.* **2018**, *75*, 170–183. [CrossRef]
32. Sasaki, K.; Nogami, S.; Fukuda, M.; Katakai, Y.; Hasegawa, A. Effect of heat treatment on bend stress relaxation of pure tungsten. *Fusion Eng. Des.* **2013**, *88*, 1735–1738. [CrossRef]
33. Yang, Y.; Gu, D.; Dai, D.; Ma, C. Laser energy absorption behavior of powder particles using ray tracing method during selective laser melting additive manufacturing of aluminum alloy. *Mater. Des.* **2018**, *143*, 12–19. [CrossRef]
34. Zhao, Y.; Koizumi, Y.; Aoyagi, K.; Yamanaka, K.; Chiba, A. Thermal properties of powder beds in energy absorption and heat transfer during additive manufacturing with electron beam. *Powder Technol.* **2021**, *381*, 44–54. [CrossRef]
35. Hu, H. Texture of metals. *Texture* **1974**, *1*, 233–258. [CrossRef]
36. Sidambe, A.T.; Tian, Y.; Prangnell, P.B.; Fox, P. Effect of processing parameters on the densification, microstructure and crystallographic texture during the laser powder bed fusion of pure tungsten. *Int. J. Refract. Met. Hard Mater.* **2019**, *78*, 254–263. [CrossRef]
37. Ren, X.; Peng, Z.; Wang, C.; Fu, Z.; Qi, L.; Miao, H. Effect of ZrC nano-powder addition on the microstructure and mechanical properties of binderless tungsten carbide fabricated by spark plasma sintering. *Int. J. Refract. Met. Hard Mater.* **2015**, *48*, 398–407. [CrossRef]

Article

Computer Simulation of Hydrodynamic and Thermal Processes in DLD Technology

Gleb A. Turichin *, Ekaterina A. Valdaytseva *, Stanislav L. Stankevich and Ilya N. Udin

World-Class Research Center, "Advanced Digital Technologies", State Marine Technical University, 190121 Saint Petersburg, Russia; s.stankevich@ilwt.smtu.ru (S.L.S.); youdin@ilwt.smtu.ru (I.N.U.)
* Correspondence: gleb@ltc.ru (G.A.T.); laser@corp.smtu.ru (E.A.V.)

Abstract: This article deals with the theoretical issues of the formation of a melt pool during the process of direct laser deposition. The shape and size of the pool depends on many parameters, such as the speed and power of the process, the optical and physical properties of the material, and the powder consumption. On the other hand, the influence of the physical processes occurring in the material on one another is significant: for instance, the heating of the powder and the substrate by laser radiation, or the formation of the free surface of the melt, taking into account the Marangoni effect. This paper proposes a model for determining the size of the melt bath, developed in a one-dimensional approximation of the boundary layer flow. The dimensions and profile of the surface and bottom of the melt pool are obtained by solving the problem of convective heat transfer. The influence of the residual temperature from the previous track, as well as the heat from the heated powder of the gas–powder jet, taking into account its spatial distribution, is considered. The simulation of the size and shape of the melt pool, as well as its free surface profile for different alloys, is performed with 316 L steel, Inconel 718 nickel alloy, and VT6 titanium alloy

Keywords: direct laser deposition; heat transfer; mass transfer; hydrodynamics; simulation of the melt pool; alloys

Citation: Turichin, G.A.; Valdaytseva, E.A.; Stankevich, S.L.; Udin, I.N. Computer Simulation of Hydrodynamic and Thermal Processes in DLD Technology. *Materials* **2021**, *14*, 4141. https://doi.org/10.3390/ma14154141

Academic Editor: Federico Mazzucato

Received: 1 July 2021
Accepted: 22 July 2021
Published: 25 July 2021

Publisher's Note: MDPI stays neutral with regard to jurisdictional claims in published maps and institutional affiliations.

Copyright: © 2021 by the authors. Licensee MDPI, Basel, Switzerland. This article is an open access article distributed under the terms and conditions of the Creative Commons Attribution (CC BY) license (https://creativecommons.org/licenses/by/4.0/).

1. Introduction

The direct laser deposition (DLD) process, according to the classification considered in [1], is currently becoming a more and more promising technology for the additive production of parts for various purposes in shipbuilding, aircraft construction, mechanical engineering, and other industries. There is already a positive experience in the manufacture of ship fittings, propellers [2,3], water jet propellers [4,5], large-sized products and machine parts [6], high-pressure vessels, and others. This technology belongs to the direct metal deposition (DMD) technologies, in which the product is formed from a metal powder supplied by a gas jet directly into the area of action of focused laser radiation. In this case, the heating and melting of the powder and the substrate is controlled by equipment in order to maintain the process in the stability zone. The high power of the laser source should ensure and maintain a high melting rate of the powder in order to ensure high process productivity up to 1.5–2.5 kg/h [4]. However, this can lead both to the appearance of instability of the wall formation described in [7], and to a deviation from the specified dimensions [8]. A large number of important parameters of the processing mode, and the danger of leaving the process in the zone of unstable formation, greatly complicate the selection of technological parameters by the experimental method. Therefore, the availability of a physically adequate and fast mathematical model that allows the performance of numerical modeling will facilitate the development and optimization of technological parameters for the direct laser deposition process. A large number of papers describe the use of various numerical schemes of finite element analysis [9–11], analytical models [12], and even statistical models [13] for modeling thermal and hydrodynamic processes. However, in most of them, the case of cladding the bead on a thick and wide substrate is considered. This is not suitable for thin walls, due to differences in boundary conditions. The influence of the Marangoni effect, capillary forces, and the mutual influence of hydrodynamics and heat transfer in the melt pool are also rarely taken into account. Finite element analysis allows scientists to take these factors into account, but the calculation time can be hours, or even days. The presented work is a development of the model developed earlier in [6,7,14,15]; it was developed specifically for modeling the process of forming thin-walled structures. This model takes into account the Marangoni effect, the transfer of the powder by a gas jet, the heating of the powder by laser radiation, the interaction of the jet with the substrate, and heat transfer in the solid and liquid phases, as well as the hydrodynamics of the melt pool. This work presents the results of theoretical studies and modeling of joint thermal and hydrodynamic processes in the stationary case for the DLD process, taking into account the influence of the heated powder on the melt pool.

2. Materials and Methods

2.1. Melt Flow Description

The high-speed DLD process is characterized by the formation of a melt pool with a length "L" much larger than the width "b" and depth "H" (Figure 1).

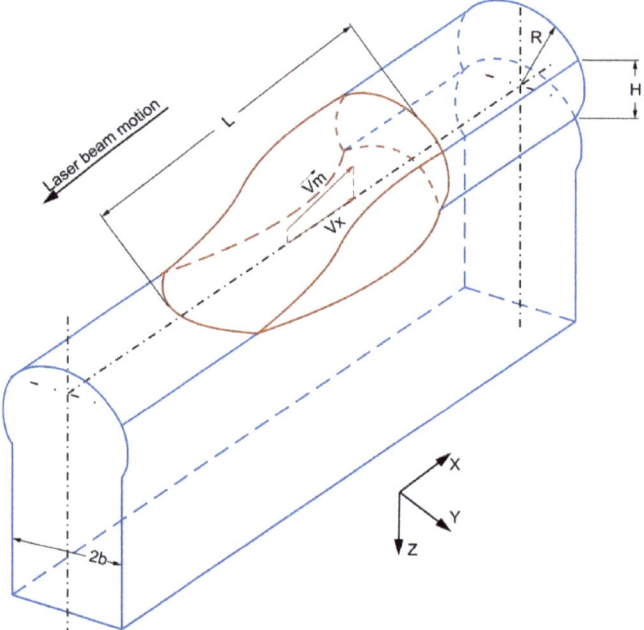

Figure 1. Design diagram of the deposited wall and the melt pool.

In this case, to describe the velocity field in the melt, we can limit ourselves to a one-dimensional formulation, and use the approximation of a one-dimensional boundary layer. In this case, the longitudinal velocity V_x, directed along the direction of movement of the laser, is much greater than the transverse velocities V_y and V_z. In the case of a steady-state process, the Navier–Stokes fluid motion equation can be written as:

$$V_x \frac{\partial V_x}{\partial x} = -\frac{1}{\rho}\frac{\partial p}{\partial x} + \nu \frac{\partial^2 V_x}{\partial z^2} \qquad (1)$$

The boundary condition at the "bottom" of the melt pool is given as:

$$V_x|_{z=0} = 0,$$

The boundary condition on the "top" surface is obtained from the requirement of the stress tensor continuity:

$$-\eta \frac{\partial V_x}{\partial z}\bigg|_{z=H} = \frac{\partial \sigma}{\partial x}$$

We assume that the temperature change along the melt pool surface is much less than the average temperature of its surface. We assume that T_s is the maximum surface temperature, while T_m and T_b are the melting and evaporation temperatures, respectively. Assuming that the law of temperature drop to the tail of the melt pool is linear, then we can write:

$$\eta \frac{\partial V_x}{\partial z}\bigg|_{z=H} = \frac{\sigma}{L}\frac{T_s - T_m}{T_b - T_m} = \frac{\sigma^*}{L} \quad (2)$$

Let the velocity distribution of the liquid in the melt have a parabolic shape. This approach will ensure that the boundary conditions and the conditions of mass flow conservation along the "x" axis are met.

$$V_x(z) = V_x\left(\alpha + \beta z + \gamma z^2\right) \quad (3)$$

Substituting Equation (3) into the boundary conditions, we obtain expressions for the coefficients of the equation:

$$\begin{aligned} \alpha &= 0 \\ \gamma &= -\frac{3}{4H}\left(\frac{2}{H} - \frac{\sigma^*}{L\eta V_x}\right) \\ \beta &= \frac{2}{H}\left(1 + \frac{H}{4}\left(\frac{2}{H} - \frac{\sigma^*}{L\eta V_x}\right)\right) = \frac{2}{H}\left(\frac{3}{2} - \frac{\sigma^*}{4\eta V_x}\frac{H}{L}\right) = \frac{3}{H} - \frac{\sigma^*}{2\eta V_x L} \end{aligned} \quad (4)$$

Then Equation (1) will look like this:

$$V_x \frac{\partial V_x}{\partial x} = -\frac{1}{\rho}\frac{\partial p}{\partial x} - 3\nu \frac{V_x}{H^2} + \frac{3\sigma^*}{2\rho L H} \quad (5)$$

We will use the continuity equation to relate the position of the weld pool surface to the melt velocity V_x. Here, it is necessary to take into account that the density of the mass flow $j(x)$ incident on the melt surface is determined by a gas–powder jet. Then the continuity equation for the flow will be written as follows:

$$\frac{\partial}{\partial x}(V_x H) = \frac{j(x)}{\rho} \quad (6)$$

Considering that the pressure in the melt:

$$p = \frac{\sigma}{R}$$

and when $H < b$ we get:

$$R \approx b + \frac{H^2}{2b},\ p \approx \frac{\sigma}{b} - \frac{\sigma H^2}{2b^3}$$

In this case, the terms of the Navier–Stokes equation associated with the change in the transverse radius of curvature of the surface will look like:

$$\frac{\partial p}{\partial x} = \frac{\sigma}{2b^3}\frac{H \partial H}{\partial x}$$

A change in the longitudinal radius of the curvature of the melt surface will give an additional pressure, which can be expressed as:

$$p_{add} = \sigma \frac{\partial^2 H}{\partial x^2}.$$

Then, one can write:

$$V_x \frac{\partial V_x}{\partial x} = \frac{\sigma}{\rho b^3} \frac{H \partial H}{\partial x} - 3\nu \frac{V_x}{H^2} + \frac{3\sigma^*}{2\rho L H} \tag{7}$$

After integration, we will get:

$$H(x) = \frac{1}{\rho v_x} \int_0^x j(x) dx \tag{8}$$

The boundary conditions for the last equation can be written as:

$$H|_{x=0} = 0, \quad \frac{\partial H}{\partial x}\bigg|_{x=0} = 0, \quad \frac{\partial H}{\partial x}\bigg|_{x=L} = 0.$$

Using Equation (8) in Equation (7), and solving it numerically, we obtain the profile of the upper surface of the melt pool, taking into account the Marangoni effect. The parameters of the length "L" and the depth "H" of the melt pool are determined by the solution to the heat transfer problem.

2.2. Influence of the Powder Jet on the Heat Transfer in the Deposited Wall

As was shown in [6], the thermal field when applying a single deposited bead to a thin wall in a steady state can be described by the equation of convective heat transfer. At the same time, since the wall width is comparable to the diameter of the laser beam spot, the temperature gradient along the y axis can be neglected.

$$V_x \frac{\partial T}{\partial x} = \chi \left(\frac{\partial^2 T}{\partial x^2} + \frac{\partial^2 T}{\partial z^2} \right) \tag{9}$$

The boundary conditions for this case can be written as:

$$-\lambda \frac{\partial T}{\partial z}\bigg|_{z=0} = q(x), \quad T|_{z \to \infty} \to T_0 \tag{10}$$

where λ and χ correspond to heat conductivity and thermal diffusivity coefficients, respectively; $q(x)$ is the distribution of the total energy flow on the melt pool surface; and T_0 is the initial temperature of the substrate.

Since the energy flux density on the pool surface includes the laser radiation intensity I and the convective heat flux brought by the heated powder, we can write:

$$q(x) = I(x) \cdot A + j(x) \cdot c \cdot (T_p(x) - T_0) \tag{11}$$

To determine T_p, we can use a well-known analytical solution to the problem of temperature distribution in a homogeneous ball of radius R with an initial temperature T_0 for the case when a constant heat flow q_p is fed into the ball through its surface [16,17]:

$$T_i(r,t) = T_0 + \frac{q_p R}{\lambda} \left(\frac{3\chi t}{R^2} - \frac{3R^2 - 5r^2}{10R^2} \right) - \frac{2q_p}{\lambda R r} \sum_{i=1}^{\infty} \frac{\sin(\mu_i r)}{\mu_i^3 \cos(R\mu_i)} e^{-\chi \mu_i^2 t}$$

where μ_n are the positive roots of the equation $tg(R\mu) = R\mu$; and t is the heating time, which is determined for each particle as the time of flight through the laser radiation zone before it enters the melt pool. Knowing the density of the powder flow in the gas–powder jet, the trajectory of the particles, their size, and the amount that got into the melt pool—for example, from [18]—it is possible to obtain the temperature distribution $T_p(x)$ in the gas—powder jet at the time of meeting with the melt surface.

Furthermore, using the method of solving the convective heat transfer equation described in [6], for the temperature field during surfacing of the i-th layer, we obtain:

$$T(x,z) = \frac{e^{-\frac{Vx}{2\chi}}}{\lambda} \int_{-\infty}^{\infty} (A \cdot I(x') + j(x')c(T_p(x') - T_0))e^{-\frac{Vx'}{2\chi}} K_0\left(\frac{V}{2\chi}\sqrt{z^2 + (x-x')^2}\right)dx' + T_w \quad (12)$$

where T_w is the residual temperature of the previous layer when applying the bead, determined by the product construction strategy.

3. Results and Discussion

The proposed model was tested at the Institute of Laser and Welding Technologies of St. Petersburg State Marine Technical University, for various products and materials. A comparison of the calculation results and experimental data showed that the error does not exceed 20%; this is a good indicator of the performance of a fast, semi-analytical model.

For the calculations, data on physical and optical properties of the works [19–26] were used. The data shown in Table 1 were averaged.

Table 1. Thermophysical and optical properties of the chosen materials.

Properties	Inconel 718	VT6	316 L
Heat capacity, J/(G·K)	0.435	0.546	0.45
Heat conductivity W/(m·K)	8.9	26	30
Density, kg/m^3	8190	4430	7800
Melting point, K	1600	1920	1710
Reflectivity for 1.06 µm, %	77	61	68

Examples of surface profile calculations for different values of motion speed and powder feed rate are shown in Figure 2. It is evident that the surface profile depends on the melt pool length "L" and depth "H".

A comparison of the melt pool surface profiles for different materials is shown in Figure 3. It can be seen from the figure that with the same mode parameters, the VT6 alloy gives a greater increment in height when the bead is deposited. At the same time, the process efficiency for steel and nickel alloy is approximately the same.

The impact of the heated powder jet on the surface temperature distribution along melt pool surface is shown in Figure 4.

Calculations show that even a small addition of energy flux with the heated powder leads to an increase in the melt pool's size (Figure 5).

For all materials, the contribution of heat from the heated powder was significant. In addition, due to the greater absorption capacity of the titanium alloy, not only was the melt pool length increased, but also the depth. For Inconel 718 and 316 L steel, this effect was not so significant. The figure shows that for 316 L steel and Inconel 718 alloy, there is a more elongated, but shallow melt pool, while for titanium alloy, on the other hand, there is a shorter and deeper melt pool. In the presence of a smaller pool, the amount of powder entering the melt will be less, and the efficiency of the process will decrease. It turns out that despite the higher absorption coefficient, having a melt pool with a smaller surface area, the efficiency of the deposition process for titanium alloy may be less than expected. This fact should be taken into account when selecting processing modes.

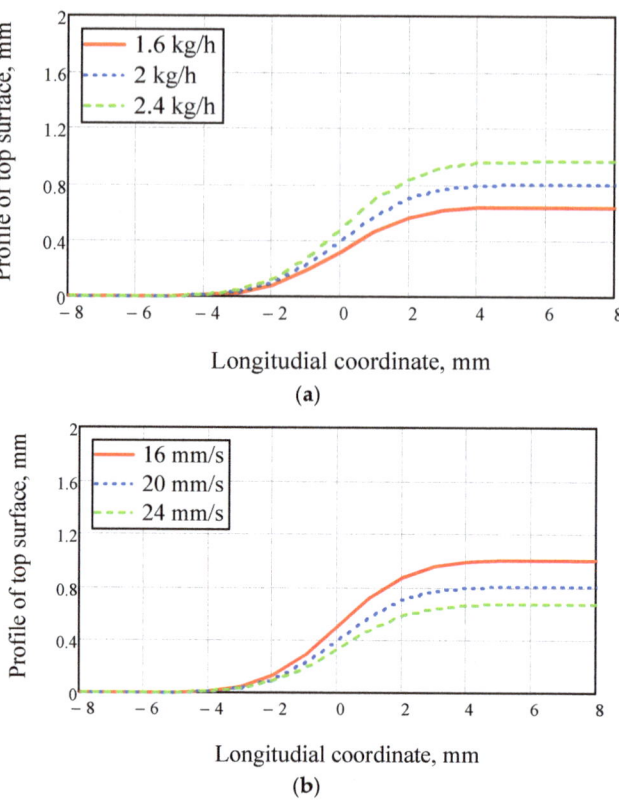

Figure 2. Melt pool top surface profile for 316 L steel for different powder feed rates (**a**) and motion speeds (**b**); laser beam radius on the surface was 2.5 mm, beam power was 2000 W, and powder jet diameter was 3 mm.

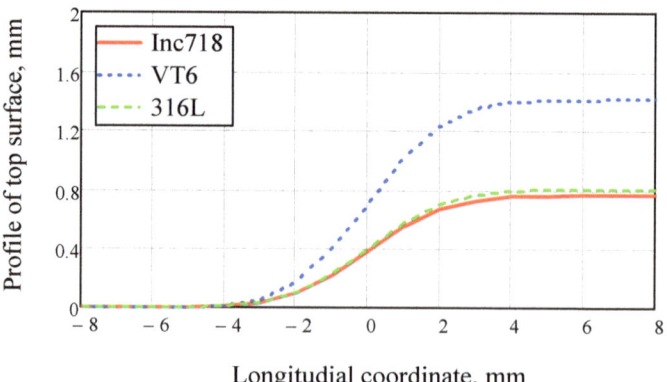

Figure 3. Melt pool top surface profiles for Inconel718, VT6, and 316 L steel in comparison with one another. Motion speed was 20 mm/s, laser beam power was 2000 W, laser beam radius on the surface was 2.5 mm, powder jet diameter was 3 mm, and powder feed rate was 2 kg/h.

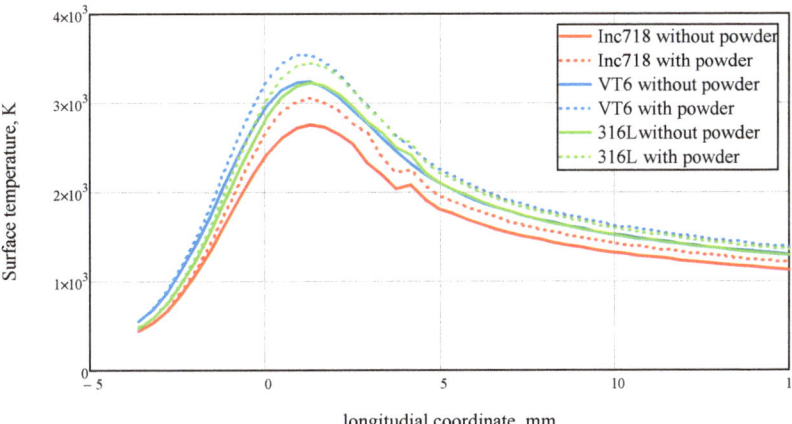

Figure 4. Temperature distribution along the melt pool length. Laser power was 2000 W, motion speed was 20 mm/s, laser beam radius on the surface was 2.5 mm, powder jet diameter was 3 mm, and powder feed rate was 2 kg/h.

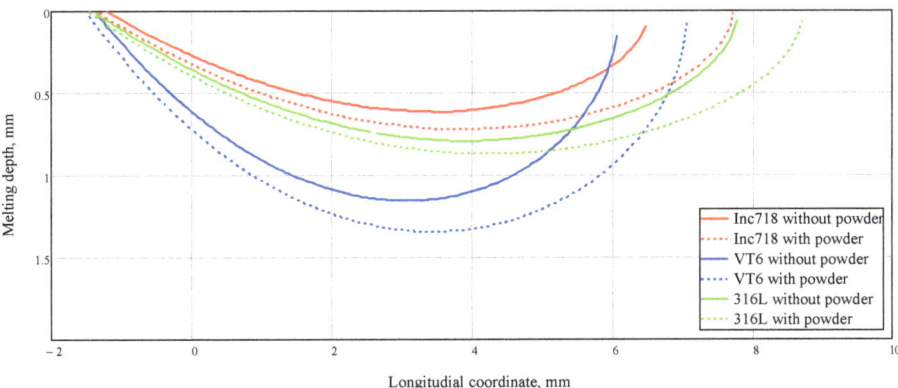

Figure 5. Melt pool shape. Laser power was 2000 W, motion speed was 20 mm/s, laser beam radius on the surface was 2.5 mm, powder jet diameter was 3 mm, and powder feed rate was 2 kg/h.

4. Conclusions

Direct laser deposition is a complex physical process. When developing new modes, a number of experiments can be replaced by a physically adequate mathematical modeling. If possible, the mutual influence of different processes on one other should be taken into account. The hydrodynamics of the melt are inextricably connected to the temperature field formed in the substrate under the action of laser radiation. In addition to the radiation itself, the final values of the temperature field in general, and the surface temperature in particular, are significantly affected by the heated powder entering the melt via a gas–powder jet. The effect of additional heat input from the powder on the pool length is noticeable for all of the materials considered—the simulation results clearly demonstrated this. Such an increase can be predicted with high probability for any metal materials. The quantitative contribution of heat from the powder depends on both the thermophysical and optical properties of the material. For some materials, it may be significant to increase not only the length of the melt pool, but also its depth—as, for example, it turned out to be for a titanium alloy. However, it should be borne in mind that reducing the size of the pool at the same mass flow density will lead to a decrease in the efficiency of the deposition process. With a significant variability in the properties of materials, the choice of parameters should mainly be carried out using mathematical modeling, since the experimental selection of modes can be extremely time-consuming.

Author Contributions: Conceptualization, G.A.T.; methodology, G.A.T. and E.A.V.; software, E.A.V.; validation, S.L.S. and I.N.U.; formal analysis, E.A.V.; investigation, G.A.T., E.A.V., S.L.S. and I.N.U.; resources, G.A.T. and E.A.V.; data curation, S.L.S. and I.N.U.; writing—original draft preparation, G.A.T. and E.A.V.; writing—review and editing, G.A.T. and E.A.V.; visualization, S.L.S. and I.N.U.; supervision, G.A.T.; project administration E.A.V.; funding acquisition, G.A.T. All authors have read and agreed to the published version of the manuscript.

Funding: This research was funded by the Ministry of Science and Higher Education of the Russian Federation as part of the World-Class Research Center Program: Advanced Digital Technologies (contract No. 075-15-2020-903 dated 16.11.2020).

Institutional Review Board Statement: Not applicable.

Informed Consent Statement: Not applicable.

Data Availability Statement: The data presented in this study are available on request from the corresponding author.

Conflicts of Interest: The authors declare no conflict of interest.

Nomenclature

L	Melt pool length, mm
H	Melt pool depth, mm
b	Melt pool half-width, mm
R	Curvature radius of a bead profile
V	Laser beam motion speed, mm/s
$V_m(x,y,z)$	Liquid metal flow velocity, mm/s
V_x, V_y, V_z	X, Y, and Z components of the liquid metal flow velocity, mm/s
c	Heat capacity, J/(kg·K)
λ	Heat conductivity, W/(m·K)
χ	Thermal diffusivity, m^2/s
ρ	Density, kg/m^3
ν	Kinematic viscosity of the liquid melt, m^2/s
η	Dynamic viscosity of the liquid melt, Pa·s
α, β, γ	Coefficients of the parabolic equation of the melt velocity X component
σ, σ^*	Surface tension, N/m
T_0	Initial temperature, environment temperature, K
T_m	Melting point, K
T_b	Boiling point, K
T_s	Maximum surface temperature, K
T_w	Residual temperature of the previous deposited bead, K
p	Pressure, Pa
p_{add}	Additional pressure over change of the longitudinal curvature radius, Pa
$j(x)$	Powder mass flow density, kg/(s·m^2)
$q(x)$	Distribution of the total energy flux on the melt pool surface, W/m
$I(x)$	Intensity of the laser beam radiation, W/m
A	Light energy absorption coefficient
$T_i(r,t)$	Temperature of a particle at co-ordinate r and heating time t, K
$T_p(x)$	Temperature distribution in the powder jet along the X axis, K

References

1. Thompson, S.M.; Bian, L.; Shamsaei, N.; Yadollahi, A. An overview of Direct Laser Deposition for additive manufacturing; Part I: Transport phenomena, modeling and diagnostics. *Addit. Manuf.* **2015**, *8*, 36–62. [CrossRef]
2. Staiano, G.; Gloria, A.; Ausanio, G.; Lanzotti, A.; Pensa, C.; Martorelli, M. Experimental study on hydrodynamic performances of naval propellers to adopt new additive manufacturing processes. *Int. J. Interact. Des. Manuf. (IJIDeM)* **2016**, *12*, 1–14. [CrossRef]
3. Ding, Y.; Dwivedi, R.; Kovacevic, R. Process planning for 8-axis robotized laser-based direct metal deposition system: A case on building revolved part. *Robot. Comput. Manuf.* **2017**, *44*, 67–76. [CrossRef]
4. Korsmik, R.; Tsybulskiy, I.; Rodionov, A.; Klimova-Korsmik, O.; Gogolukhina, M.; Ivanov, S.; Zadykyan, G.; Mendagaliev, R. The approaches to design and manufacturing of large-sized marine machinery parts by direct laser deposition. *Procedia CIRP* **2020**, *94*, 298–303. [CrossRef]
5. Mendagaliev, R.; Klimova-Korsmik, O.; Promakhov, V.; Schulz, N.; Zhukov, A.; Klimenko, V.; Olisov, A. Heat Treatment of Corrosion Resistant Steel for Water Propellers Fabricated by Direct Laser Deposition. *Materials* **2020**, *13*, 2738. [CrossRef] [PubMed]
6. Turichin, G.; Klimova-Korsmik, O.; Babkin, K.; Ivanov, S.Y. Additive manufacturing of large parts. *Addit. Manuf.* **2021**, 531–568. [CrossRef]
7. Turichin, G.; Zemlyakov, E.; Klimova, O.; Babkin, K. Hydrodynamic Instability in High-speed Direct Laser Deposition for Additive Manufacturing. *Phys. Procedia* **2016**, *83*, 674–683. [CrossRef]
8. Turichin, G.; Zemlyakov, E.; Babkin, K.; Ivanov, S.; Vildanov, A. Analysis of distortion during laser metal deposition of large parts. *Procedia CIRP* **2018**, *74*, 154–157. [CrossRef]
9. Fu, G.; Zhang, D.Z.; He, A.N.; Mao, Z.; Zhang, K. Finite Element Analysis of Interaction of Laser Beam with Material in Laser Metal Powder Bed Fusion Process. *Materials* **2018**, *11*, 765. [CrossRef] [PubMed]
10. Razavykia, A.; Brusa, E.; Delprete, C.; Yavari, R. An Overview of Additive Manufacturing Technologies—A Review to Technical Synthesis in Numerical Study of Selective Laser Melting. *Materials* **2020**, *13*, 3895. [CrossRef]
11. Guo, Z.; Wang, L.; Wang, C.; Ding, X.; Liu, J. Heat Transfer, Molten Pool Flow Micro-Simulation, and Experimental Research on Molybdenum Alloys Fabricated via Selective Laser Melting. *Materials* **2020**, *14*, 75. [CrossRef] [PubMed]
12. Ning, J.; Sievers, D.E.; Garmestani, H.; Liang, S.Y. Analytical Thermal Modeling of Metal Additive Manufacturing by Heat Sink Solution. *Materials* **2019**, *12*, 2568. [CrossRef]
13. Caiazzo, F.; Alfieri, V. Simulation of Laser-assisted Directed Energy Deposition of Aluminum Powder: Prediction of Geometry and Temperature Evolution. *Materials* **2019**, *12*, 2100. [CrossRef] [PubMed]
14. Turichin, G.; Valdaytseva, E.; Pozdeeva, E.; Zemlykov, E. Influence of aerodynamic force on powder transfer to flat substrate in laser cladding, BTLA. In Proceedings of the Six International Scientific and Technical Conference, Saint Petersburg, Russia, 2–6 September 2009; pp. 42–47.
15. Turichin, G.A.; Somonov, V.V.; Klimova, O.G. Investigation and Modeling of the Process of Formation of the Pad Weld and its Microstructure during Laser Cladding by Radiation of High Power Fiber Laser. *Appl. Mech. Mater.* **2014**, *682*, 160–165. [CrossRef]
16. Bitzadze, A.V.; Kalinichenko, D.F. *Collection of Problems on Equations of Mathematical Physics*; Nauka: Moscow, Russia, 1986; p. 224.
17. Stankevich, S.L.; Korsmik, R.S.; Valdaytseva, E.A. Modeling of bead formation process during laser cladding. *J. Phys. Conf. Ser.* **2017**, *857*, 12045. [CrossRef]
18. Stankevich, S.L.; Topalov, I.K.; Golovin, P.A.; Valdaytseva, E.A. Study of metallic powder flow in discrete coaxial nozzles. *J. Phys. Conf. Ser.* **2018**, *1109*, 012008. [CrossRef]
19. Johnson, P.B.; Christy, R.W. Optical constants of transition metals: Ti, V, Cr, Mn, Fe, Co, Ni, and Pd. *Phys. Rev. B* **1974**, *9*, 5056–5070. [CrossRef]
20. Werner, W.S.M.; Glantschnig, K.; Draxl, C. Optical Constants and Inelastic Electron-Scattering Data for 17 Elemental Metals. *J. Phys. Chem. Ref. Data* **2009**, *38*, 1013–1092. [CrossRef]
21. Ordal, M.A.; Bell, R.J.; Alexander, R.W.; Long, L.L.; Querry, M.R. Optical properties of Au, Ni, and Pb at submillimeter wavelengths. *Appl. Opt.* **1987**, *26*, 744–752. [CrossRef]
22. Rakić, A.D.; Djurišić, A.B.; Elazar, J.M.; Majewski, M.L. Optical properties of metallic films for vertical-cavity optoelectronic devices. *Appl. Opt.* **1998**, *37*, 5271–5283. [CrossRef] [PubMed]
23. Querry, M.R. *Optical Constants*; Contra University of Missouri-Kansas City: Kansas City, MO, USA, 1985.
24. Palm, K.; Murray, J.; Narayan, T.C.; Munday, J.N. Dynamic Optical Properties of Metal Hydrides. *Acs Photonics* **2018**, *5*, 4677–4686. [CrossRef]
25. Xie, J.; Kar, A. Laser welding of thin sheet steel with surface oxidation. *Weld. J.* **1999**, *78*, 343-s.
26. Juan, J.; Peter, N. Thermophysical Properties. *ASM Handb.* **2008**, *15*, 468–481.

Article

Features of Heat Treatment the Ti-6Al-4V GTD Blades Manufactured by DLD Additive Technology

Marina Gushchina [1,*], Gleb Turichin [1], Olga Klimova-Korsmik [1], Konstantin Babkin [1] and Lyubov Maggeramova [2]

1. World-Class Research Center "Advanced Digital Technologies", State Marine Technical University, 190121 Saint-Petersburg, Russia; gleb@ltc.ru (G.T.); o.klimova@ltc.ru (O.K.-K.); babkin@ilwt.smtu.ru (K.B.)
2. Central Institute of Aviation Motors (CIAM), 111116 Moscow, Russia; mag@ciam.ru
* Correspondence: gushcina_mo@corp.smtu.ru

Abstract: Additive manufacturing of titanium alloys is one of the fastest growing areas of 3D metal printing. The use of AM methods for parts production in the aviation industry is especially promising. During the deposition of products with differently sized cross-sections, the thermal history changes, which leads to non-uniformity of the structure and properties. Such heterogeneity can lead to failure of the product during operation. The structure of deposited parts, depending on the thermal cycle, may consist of α', $\alpha + \alpha' + \beta'$, and $\alpha + \beta$ in different ratios. This problem can be solved by using heat treatment (HT). This paper presents research aimed towards the determination of optimal heat treatment parameters that allows the reception of the uniform formation of properties in the after-treatment state, regardless of the initial structure and properties, using the example of a deposited Ti-6Al-4V gas turbine blade.

Keywords: Ti-6Al-4V; direct energy deposition; thermal history; annealing; phase composition; microstructure; tensile properties

1. Introduction

The transition of the world aircraft industry to innovative technologies, including the replacement of metal structures with composite materials, the development of additive manufacturing, and the introduction of new artificial intelligence systems in the aircraft control system, is becoming an increasingly relevant trend.

Computer engineering is widely used to create new materials in the aviation industry, which reduces the cost of the production of expensive full-scale prototypes by using virtual models. Additive technologies, most commonly based on the use of virtual models, are widely available due to the fact that they allow the manufacturing of products with complex geometric shapes and profiles. The use of such advanced technologies will significantly reduce the time introduction of products to the market and their cost, reduce material consumption, and reduce the product failure, which is a clear advantage for use in most industries [1–3].

The aviation industry is characterized by increased requirements for structural materials, and in some cases is a main customer and consumer of new materials and technologies [4]. A range of AM techniques are now available. The following additive manufacturing methods have become particularly popular in the aircraft industry for metal parts: selective laser melting (SLM), electron beam melting (EBM), and direct laser deposition (DLD). In SLM and EBM, thin powder layers are consolidated layer by layer using electron or laser beam scanning, and a layer is formed along the path of the corresponding user-created model [5,6]. In DLD, the formation of a layer occurs by the coaxial supply of laser radiation and powder through special nozzles. Each of the methods has its advantages and disadvantages [7]. Due to the design features of the machines, SLM and EBM are more suitable for small- and medium-sized products, while DLD is attractive due to the possibility of producing large-sized products [7,8].

DLD is applicable to producing products from alloys based on iron, titanium, and nickel, as well as composite materials or compositionally graded materials. The α + β titanium alloy Ti-6Al-4V is widely used in aerospace applications, and much research has been conducted on AM with this alloy. There are many works devoted to the study of the structure and properties of additively manufactured Ti-6Al-4V in the building as well as the post-processing state [9–12]. Heat treatments (HTs) are solved problems of structural and phase inhomogeneity, residual stresses, and anisotropy. Hot isostatic pressing (HIP) of deposited samples allows the removal of defects, such as pores and non-fusion [13].

The fundamental possibility of achieving high strength and fatigue properties of model samples deposited from Ti-6Al-4V powder is shown in a number of works [14–18]. However, in most studies, the represented data for model samples have the form of plates, which are far from real parts. For example, [19,20] presented an investigation regarding the influence of product shape and thickness on other properties, and it was found that the shape has a more significant effect on thin-walled samples than on thick-walled samples. The overall dimensions of the samples also affected the properties: samples with a thick wall had higher strength, which, according to the authors, is associated with the size and morphology of the initial β-grain [21,22]. Thus, we can assume that the combination of thin-walled and thick-walled elements in one part can lead to significant heterogeneity of properties and structures throughout the product, which can negatively affect the performance of the entire part.

In this particular case, this problem can be solved by careful and accurate selection of modes, which requires a significant amount of work. On a global scale, the problem can be solved by developing the optimal heat treatment parameters that, regardless of the initial phase composition and grain size, the uniform structure and properties can provide.

This paper presents studies on the developing and selection of optimal heat treatment modes that ensure the uniform formation of properties in the deposited compressor gas turbine engine (GTE) blades of a Ti-6Al-4V after heat treatment, regardless of the initial structure and properties. In this work, we investigated a wide range of HT temperatures for laser deposited Ti-6Al-4V to establish the relationships between heating temperature and the mechanical properties. The selected mode was tested for samples with different initial microstructures and phase compositions.

2. Materials and Methods

All the compressor gas turbine engine blades investigated in this research were built using the Ti-6Al-4V titanium alloy powder, with a fraction of 45–90 microns produced by PREP (plasma rotating electrode process). Samples were produced by the robotic complex developed at St. Petersburg State Marine University. The complex includes an LS-5 fiber laser, an anthropomorphic robot, a 6 m^3 protecting chamber, a two-axis positioner, a powder feeder, and a control system. The path was generated in the Additive Control 1.0 program (ILWT, Saint Petersburg, Russia) using a 3D model of the blade. Figure 1a shows the trajectory of the deposition of the blade. The technological parameters for a Ti-6Al-4V alloy are presented in Table 1. The process of blade deposition and the final result are shown in Figure 1b,c. The process was carried out in a chamber with an argon atmosphere, and the oxygen level in the chamber was 2000 ppm.

Table 1. Technological parameters of the Ti-6Al-4V DLD process.

Power, W	Speed, mm/s	Gas Consumption L/min	Beam Diameter, mm	Powder Consumption, %
1500	30	15	2.8	35–40

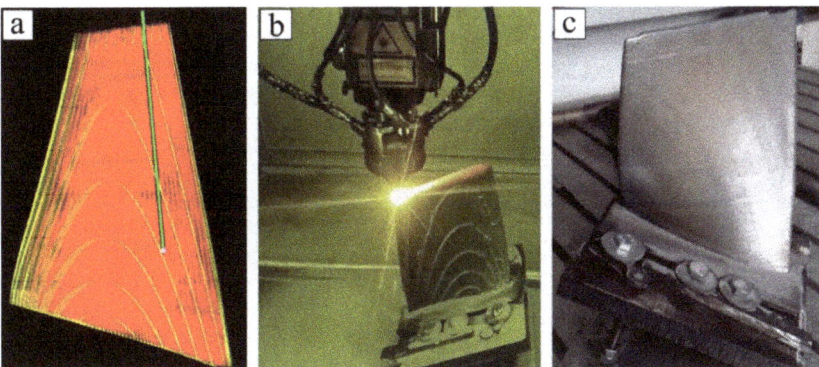

Figure 1. The blades of a gas turbine compressor; (**a**) the path generated in Additive Control; (**b**) the process of blade deposition; (**c**) for a Ti-Al-4V deposited blade.

The effect of heat treatment on the structural change was studied on the $5 \times 5 \times 10$ mm^3 pieces, which were cut from the central part of the deposited blades. The optimal heat treatment conditions for the Ti-6Al-4V titanium alloy were determined by varying the furnace heating temperature and holding time. The temperature varied within the range of 600–950 °C. The hold time was 2 h.

An air atmosphere was chosen for heat treatment in an SNOL furnace, without additional argon protection. This choice was made on the basis of preliminary studies conducted by the authors of this article. It was shown that at a temperature HT of 900 °C and a holding time of 4 h, the maximum thickness of the alpha case layer in Ti-6Al-4V, which consists of TiO$_2$ and a Ti (O) solid solution, does not exceed 200 µm (Figure 2a).

At temperatures above 600 °C, titanium actively interacts with oxygen, and forms a solid solution of up to 10 at.% in α-Ti. The diffusion rate of oxygen atoms increases with increasing temperature and holding time, but the formation of TiO$_2$ oxide on the surface decreases the diffusion rate. Thus, the effect of oxygen on the properties was insignificant when we used intervals of temperatures and time considered in the article for Ti-6Al-4V heat treatment. The assessment was based on hardness changes (Figure 2b).

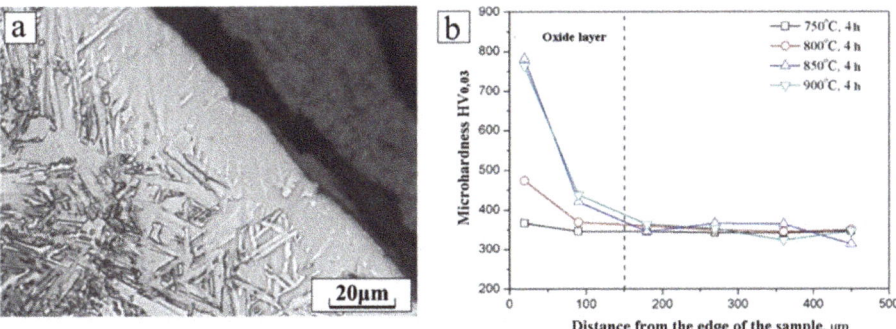

Figure 2. The formation of an oxide layer in deposited Ti-6Al-4V during heat treatment in an air atmosphere; (**a**) an oxide layer of Ti-6Al-4V after HT at 900 °C for 4 h; (**b**) the dependence of the oxide layer hardness of the holding temperature.

Plates with a size of $75 \times 15 \times 35$ mm^3 for mechanical tests were deposited in accordance with the thermal cycles corresponding to blade deposition. For the definition of the mechanical characteristics of deposited products, uniaxial tensile tests were performed. Mechanical tests were performed on a universal testing machine, the Zwick/Roell Z250

Allround series (Zwick/Roell, Ulm-Einsingen, Germany). The standard cylindrical samples were cut from the printed parts according to geometry from the ASTM E8, with a gauge diameter of 6.0 mm and a gauge length of 24.0 mm. The sample displacement was recorded using an extensometer. The optical microscope DMI 5000 (Leica, Wetzlar, Germany) with the Axalit software (Axalit, Moscow, Russia) was used for microstructural analysis. The sample surfaces were polished with grit SiC papers up to 2500 grits, with further polishing by an aluminum oxide suspension of 1 μm and final polishing with colloidal silica. As a final step, the etching procedure was conducted. A solution of 93 mL of H2O + 2 mL of HF + 5 mL of HNO3 was applied to the polished sample surfaces for 40 s.

A Bruker Advance D8 diffractometer (Bruker, Billerica, USA) with CuKα radiation (wavelength = 1.5418 Å) was used to perform the XRD analysis. The detector was a LynxEye linear position-sensitive detector (PSD) with a capture angle of 3.2 degrees 2θ. The cross-sectional surface of the samples was polished with a grit SiC paper of 2500 grits, and, after etching, measured in the 2θ range of 30°–60° with a step size of 0.05 and an incremental time of 0.02.

Microhardness measurements were performed with an FM-310 microhardness tester (Future Tech, Kawasaki, Japan). The sample surface for testing was polished with 500 and 2500 grit SiC abrasive papers. Microhardness was measured on the central part of the deposited blades.

3. Results and Discussion

The recrystallization temperature range for the Ti-6Al-4V alloy was 850–950 °C. Including this, temperatures of 850, 900, and 950 °C were investigated for heat treatment deposited samples. The pre-crystallization annealing modes were also used, since these temperatures may be sufficient to relieve stresses and metastable phase decomposition.

At temperatures above 1000 °C, recrystallization and significant growth of alpha plates occurs. This leads to a decrease in the mechanical properties for AM samples; therefore, heat treatment above 1000 °C was not considered in this work [23]. The effect of the cooling rate on the microstructure after heat treatment was also not considered, since, in previous works, it was found that the cooling rate during HT does not have a significant effect [24]. Due to the very fine martensite, the kinetics were completely different compared to treatment of equiaxed or heavily deformed microstructures. Consequently, the application of standard heat treatments shows that these treatments do not lead to the usual or expected results [24].

Before 700 °C, a layer structure was still observed, which is typical for as-deposited samples. Annealing for 2 h at a temperature of 700 °C led to the formation of new equiaxed grain near the boundaries between the layers (suggesting that, in these areas, the greatest deformation of the primary grain occurred). Thus, a shift in the temperature of the onset of the recrystallization process was observed.

This may be due to a high level of initial grain deformation due to high internal stresses (Figure 3a). Formation at the boundaries between the layers of secondary grains was also observed upon a heating temperature of 800 °C and a holding time of 2 h (Figure 3b). As the temperature rose to 900 °C, a more intense recrystallization process occurred, and the formation of new equiaxed grains occurred not only at the boundaries between the layers, but also inside the layer (Figure 3c,d). From the deformed grains, the growth of secondary under-formed grains was observed. With increasing holding time, the size of the secondary grains increased at 800–900 °C. Inside the grain, the morphology of the α-plates changed.

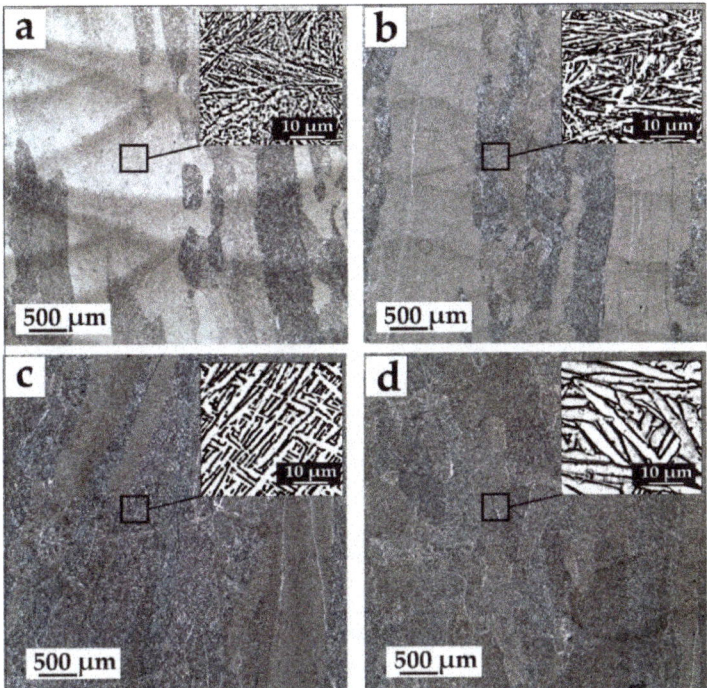

Figure 3. Microstructure of deposited Ti-6Al-4V samples after heat treatment (50× magnification): (**a**) 700 °C, (**b**) 850 °C, (**c**) 900 °C, 2 h, (**d**) 950 °C, 2 h.

The temperature of 950 °C was the boundary of the end of the recrystallization process (for α + β titanium alloys produced by conventional methods [25]). At this temperature, a morphology of the α and β plates inside the grain changed (Figure 3a). In addition, due to the closer temperature of the allotropic transformation, partial recrystallization occurred due to the transition from the low-temperature α to the high-temperature β-phase, and vice versa, upon cooling in the air. The anisotropy of the alloy decreased because of the change in grain size.

In the process of heating to 600 and 650 °C and holding for 2 h, only residual stresses were removed, there were no visible structural changes, and the form of the α/α' phase was the same, with a characteristic needle structure. The absence of the α' decomposition was also evidenced by the high hardness of the samples similar to the initial state. Due to the diffusion process, increasing to a temperature to 700 °C led to the decomposition of the metastable phase, and a thickness increase of the α phase lamellae began (Figure 3a). The morphology also changed slightly; the needle structure was replaced by a lamellar structure.

A heat treatment temperature of 800 °C with a holding time of 2 h led to the secondary α phase, which began to form along the grain boundaries, and the needles transformed into the plates (Figure 3b). There were both long and short plates located perpendicular to each other, as well as in separate packages of parallel plates.

During heat treatment with a temperature of 850 °C for 2 h, the secondary α phase was formed, and the metastable α' phase was decomposed with the formation of stable composition a + β phases. The width plates increased, and the amount of secondary α increased as well (Figure 3c). At a temperature of 900 °C and a holding time of 2 h, the structural components grew. At an annealing temperature of 950 °C, partial recrystallization occurred due to allotropic transformation, and a structure of the "basket weaving" type was formed. The thickness of the plates increased significantly in comparison with the annealing temperature of 850 °C.

At low annealing temperatures (below 600 °C), the decomposition of a' was incomplete, as is shown by the low value of the hardness.

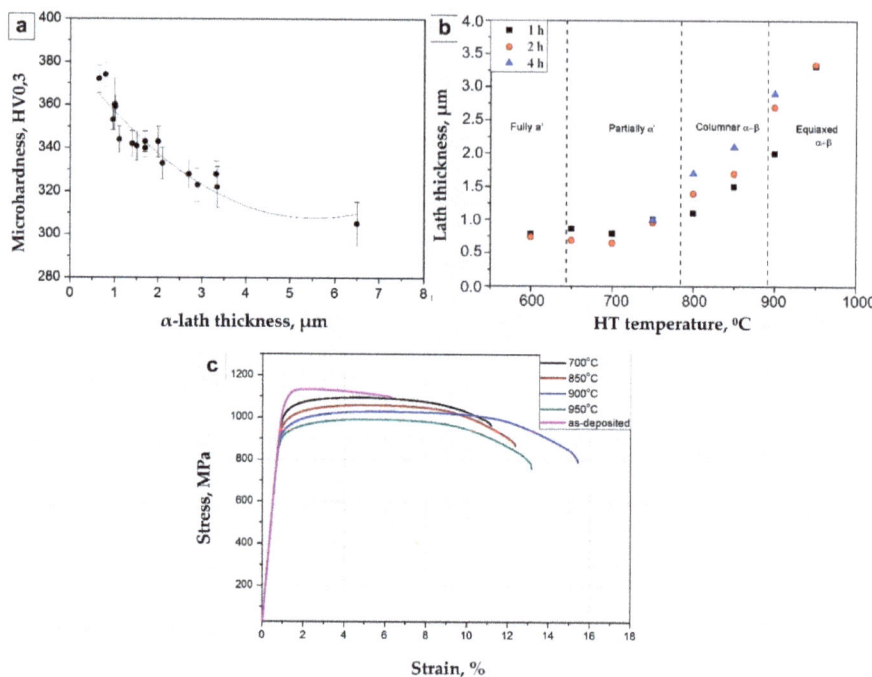

Figure 4. Experimental microhardness measurements for different α-lath thickness values; the tendency line was plotted using polynomial regression analyses (**a**); experimental data of α-lath thickness versus an aging time graph for different temperatures (**b**); typical stress–strain curves of DLD with the Ti-6Al-4V alloy performed with different HT modes (**c**).

3.1. Heat Treatment Temperature Effect on Mechanical Properties

Mechanical test results are presented in Table 2. It can be seen that temperature and time increasing led to YS reduction. On the contrary, an increase in elongation occurred up to a temperature of 900 °C; above this temperature, with a holding time of 2 h, a decrease in the relative elongation occurred. H. Galarraga et al. presented the influence of heat treatment by using different modes that consisted of several stages of electron beam-melted Ti-6Al-4V. However, for DLD, it is possible to obtain high mechanical properties using a one-stage heat treatment [12].

Table 2. The results of mechanical testing of deposited Ti-6Al-4V cylindrical samples with different HT modes.

HT	Yield Strength, MPa	Tensile Strength, MPa	Elongation, %
as-deposited	1082	1135	5.3
700 °C/AC/2 h	1019	1100	8.27
850 °C/AC/2 h	963	1055	12.2
900 °C/AC/2 h	925	1026	14.2
950 °C/AC/2 h	903.2	990.7	13

The graph of Figure 4a shows the dependence of the plates' α/α' thickness on temperature and holding time during heat treatment. As expected, an increase in the temperature and holding time led to an increase in the structure components size for the Ti-6Al-4V alloy. In addition, since, at a holding temperature of 600 °C, the hardness does not change and corresponds to the as-deposition state, the initial phase composition α + α' + β was probably retained. At temperatures of 700–750 °C, even when the holding time was 4 h, only a partial transition of α' to equilibrium α + β occurred. Starting from temperatures of 800 °C, a more intense decomposition was observed, and above 850 °C, the growth of the structural components already had a prevailing effect on microhardness.

Based on the experimental data, the Hall–Petch relationship was plotted for the deposited titanium alloy Ti-6Al-4V using polynomial regression analyses that corresponded to the data presented in [12].

The microhardness gradually decreased with an increase in the α phase plate size (Figure 4a). The microhardness measurement confirmed the decomposition of the metastable α' phase, beginning at the temperature of 700 °C. A further decrease in hardness with increasing temperature and holding time was associated with an increase of α lamellae size (Figure 4). The Hall–Petch relationship can also be observed for temperature and time variation in Figure 4b. The results of the α lath thickness measurement for all the aging temperatures and times are plotted in Figure 4b. The graph shows that α lath size grew with temperature and time. The lamellae coarsening increased with temperature. The results for various temperatures correlated with the data obtained for the SLM-ed Ti-6Al-4V alloy [26].

Analysis of the fracture surface showed that an increase in the HT temperature led to a facet size increase, which also corresponded to an increase in ductility and a decrease in tensile strength. Heat treatment at a temperature of 700 °C did not significantly affect the change in the structure; the fracture of this sample was more similar to the fracture of the as-deposited samples, also characterized by inter-crystalline fracture (Figure 5a,b,e,f). The use of heat treatment temperatures above 800 °C led to a partial or complete decomposition of metastable structures. Above 900 °C, the growth of structural components occurred, which also affected the facet size on the fracture surface (Figure 5c,d,g,h).

Figure 5. Fracture surface of as-deposited Ti-6Al-4V tensile samples performed in the horizontal direction (x-axis) for (a,e) as-deposited; (b,f) HT = 700 °C; (c,g) HT = 850 °C; and (d,h) HT = 900 °C.

A comparative analysis of the diffraction patterns of the as-deposited sample and after heat treatment using different temperatures allowed the determination of the features of the change in the intensity phases' diffraction lines. As-deposited samples had a slight shift of lines that, as noted previously, indicated the formation of a metastable α' phase due to dissolving more alloying elements in its lattice, which explained the shift [27] (Figure 6). The metastable α' was partially present in the Ti-6Al-4V up to 900 °C, and diffraction peaks of the α phase had some shifts. Complete α' phase decomposition was observed above 900 °C. As-deposited XRD-patterns also had peak broadening that indicated internal stresses. After heat treatment, the internal stress level decreased, and the peak shape became narrower. This corresponded to the data presented in the work of T. Ungar [28]. In addition, a change of (101)α and (200)α intensity could be traced. Shunyu Liu et al. associated the intensity of alpha planes variation with decomposition of martensite and the preferred grain orientation changing [29]. Moreover, corresponding with optical micrograps, Figure 4 shows that the growth of structural components is observed with increasing HT temperature. The increasing of alpha plates also influenced peak intensity.

The diffraction line intensity of the β-phase increase for deposited Ti-6Al-4V samples after heat treatment. In addition, for all heat treatments, the β-phase diffraction lines were broadened. Based on this, it can be concluded that the β-phase had some inhomogeneity in composition and stress level, which was typical for the phase in which decomposition occurred. The position of the β-phase lines was not constant, and indicate a slightly different degree of decomposition depending on the holding temperature. This was especially pronounced on the sample after heat treatment with a holding temperature of 850 °C. This indicated that at a given temperature, the holding time of 2 h was not enough to complete the phase transition processes.

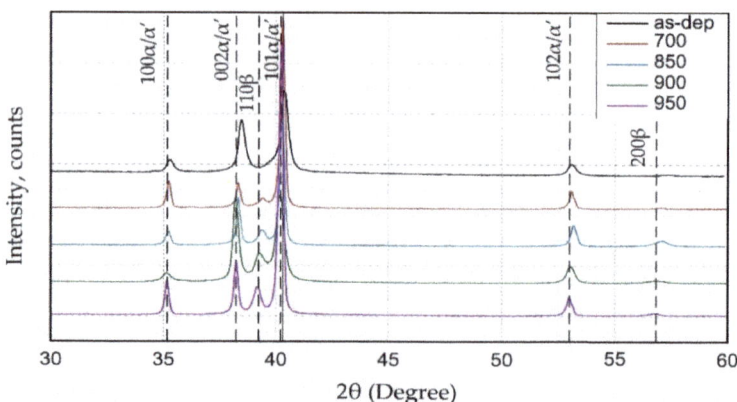

Figure 6. XRD patterns of as-deposited Ti-6Al-4V alloy and after heat treatment.

3.2. Heat Treatment of Depositede Ti-6Al-4V with Different Structures

In the deposit samples, depending on thermal cycles, heating and cooling rates, and modes parameters, different types of structures typical for Ti-6Al-4V can be observed. For different types of structures, there can be different ratios of phases: α + β, α + β, and α + α' + β. The size, distance from the substrate, and the thermal cycle, as previously indicated for various AM technologies of Ti-6Al-4V, had a significant effect on the final structure, phase composition, and the distribution of alloying elements. [30,31].

To study the effect of the selected heat treatment mode on three types of initial Ti-6Al-4V structures, samples deposited under different conditions were tested (Table 3). Changes in microstructure and properties depending on conditions have been shown in previous work [32]. It was shown that higher cooling rates were observed at the bottom of the blade near the substrate, which led to the partial or full α' formation. Above the 40th layer, heat accumulated and an equilibrium structure α + β was formed. In accordance with the above, the selected heat treatment mode was tested on three samples to show good applicability, regardless of the initial structure (Figure 7, Table 3).

Samples with nonuniform structure have different mechanical properties (Table 4). The heat treatment parameters must be selected in such a way that the properties in the initial state do not decrease if they have the required level comparable to the mechanical properties of this alloy in the rolled condition, but, at the same time, so that they can be improved if the required level is not achieved.

Table 3. Samples for testing selected heat treatment mode.

Name	Phase Composition	Deposited Blade Part
Sample 1	A + α'	Near substrate
Sample 2	α + β	Until 40th layer
Sample 3	A + α' + β	After 40th layer

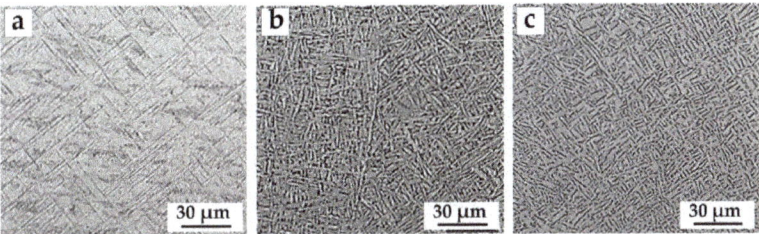

Figure 7. Microstructure of deposited Ti-6Al-4V with different initial structure (a) sample 1 (α + α'), (b) sample 2 (α + β), and (c) sample 3 (α + α' + β).

Table 4. Mechanical properties of samples deposited with different thermal cycles before and after HT.

№ of Sample	Yield Strength, MPa	Tensile Strength, MPa	Elongation, %	Microhardness HV0.03
sample 1	1082	1135	5.3	385 ± 7.98
sample 1 + HT	888	951	9.7	345 ± 8.61
sample 2	983	1047	10.3	354 ± 5.15
sample 2 + HT	899	960	10,4	339 ± 8.28
sample 3	931	999	5.8	360 ± 6.69
sample 3 + HT	895	958	10.0	341 ± 5.89

The Table 3 shows that heat treatment led to a similar plasticity level for deposited Ti-6Al-4V samples after HT, although, before there was a significant difference in properties, which was most likely associated with the ratio of equilibrium and non-equilibrium phases in the alloy.

4. Conclusions

During a direct laser deposition process from a Ti-6Al-4V powder alloy of aviation parts, which have variable geometric dimensions of the sections due to the thermal cycle's changes, a non-uniform structure can form that decreases the mechanical properties. As a result, the mechanical properties of the part become heterogeneous, which can lead to premature failure during deposition or decrease working properties. In this paper, the optimal heat treatment mode is selected, which, regardless of the structure, gives a good result and evens out the mechanical properties in the part.

(a) The stabilization of the α' phase had a strong influence on its decomposition and grain growth during subsequent heat treatment. Lamellae size showed some relation with heating temperature, illustrating the significance of the initial microstructure and heating temperature applied in post-DLD heat treatment. Above the heating temperature 850 °C, α lamella width growth became more intensive (>2.5 μm).

(b) The metastable α' is partially present in the Ti-6Al-4V up to 900 °C. Complete α' phase decomposition is observed above 900 °C. XRD results show that the β phase also has some inhomogeneity in composition and stress level that is typical for the phase in which decomposition occurs. Full-phase transformation for DLD-ed Ti-6Al-4V alloy when the temperature of heat treatment is above 900 °C occurs.

(c) An elongation increase an decrease in the tensile strength occurs with a growth in the holding temperature from 700 °C (σ_t = 1100 MPa, el = 8.27%) to 900 °C (σ_t = 1026 MPa, el = 14.2%). Above 900 °C, a decrease in elongation begins with a simultaneous decrease of tensile strength (σ_t = 990 MPa, el = 13% for 950 °C) that is associated with an increase in the α lamellae and beta grain size. The results for all properties are well above ASTM standards for forged (ASTM F1472) and cast Ti6-Al-4V (ASTM F1108).

The best mechanical characteristics of laser-deposited Ti-6Al-4V are ensured by an HT with a temperature of 900 °C and a holding time of 2 h. The selected HT parameters allow the homogeneity of properties and microstructure in DLD-ed Ti-6Al-4V parts with variable thickness and complexity, particularly in deposited GTA blades.

Author Contributions: Conceptualization, G.T. and O.K.-K.; Methodology, K.B.; formal analysis, M.G.; investigation, M.G. and O.K.-K.; resources, G.T. and L.M.; data curation, G.T., O.K.-K. and G.T.; writing—original draft preparation, M.G.; writing—review and editing, O.K.-K. and L.M.; visualization M.G, supervision, O.K.-K. and K.B.; project administration, G.T.; funding acquisition, G.T. All authors have read and agreed to the published version of the manuscript.

Funding: This research is funded by the Ministry of Science and Higher Education of the Russian Federation as part of the World-class Research Center program: Advanced Digital Technologies (contract No. 075-15-2020-903 dated 16.11.2020).

Institutional Review Board Statement: Not applicable.

Informed Consent Statement: Not applicable.

Conflicts of Interest: The authors declare no conflict of interest.

References

1. Thompson, S.M.; Bian, L.; Shamsaei, N.; Yadollahi, A. An overview of Direct Laser Deposition for additive manufacturing; Part I: Transport phenomena, modeling and diagnostics. *Addit. Manuf.* **2015**, *8*, 36–62. [CrossRef]
2. Mendagaliyev, R.; Turichin, G.; Klimova-Korsmik, O.; Zotov, O.; Eremeev, A. Microstructure and Mechanical Properties of Laser Metal Deposited Cold-Resistant Steel for Arctic Application. *Procedia Manuf.* **2019**, *36*, 249–255. [CrossRef]
3. Buczak, N.; Hassel, T.; Kislov, N.G.; Klimova-Korsmik, O.G.; Turichin, G.A.; Magerramova, L.A. Phase and Structural Transformations in Heat Resistant Alloys during Direct Laser Deposition. *Key Eng. Mater.* **2019**, *822*, 389–395. [CrossRef]
4. Shapiro, A.; Borgonia, J.P.; Chen, Q.N.; Dillon, R.P.; McEnerney, B.; Polit-Casillas, R.; Soloway, L. Additive Manufacturing for Aerospace Flight Applications. *J. Spacecr. Rocket.* **2016**, *53*, 952–959. [CrossRef]
5. Singh, D.D.; Mahender, T.; Reddy, A.R. Powder bed fusion process: A brief review. *Mater. Today Proc.* **2021**, *46*, 350–355. [CrossRef]
6. Khorasani, A.M.; Gibson, I.; Ghasemi, A.; Ghaderi, A. A comprehensive study on variability of relative density in selective laser melting of Ti-6Al-4V. *Virtual Phys. Prototyp.* **2019**, *14*, 349–359. [CrossRef]
7. Liu, S.; Shin, Y.C. Additive manufacturing of Ti6Al4V alloy: A review. *Mater. Des.* **2019**, *164*, 107552. [CrossRef]
8. Kelbassa, I.; Wohlers, T.; Caffrey, T. Quo vadis, laser additive manufacturing? *J. Laser Appl.* **2012**, *24*, 50101. [CrossRef]
9. Antonysamy, A.; Meyer, J.; Prangnell, P. Effect of build geometry on the β-grain structure and texture in additive manufacture of Ti6Al4V by selective electron beam melting. *Mater. Charact.* **2013**, *84*, 153–168. [CrossRef]
10. Shalnova, S.; Panova, G., Buczak, N. Structure and Phase Composition of Ti-6Al-4V Samples Produced by Direct Laser Deposition. *Key Eng. Mater.* **2019**, *822*, 467–472. [CrossRef]
11. Carroll, B.E.; Palmer, T.A.; Beese, A. Anisotropic tensile behavior of Ti–6Al–4V components fabricated with directed energy deposition additive manufacturing. *Acta Mater.* **2015**, *87*, 309–320. [CrossRef]
12. Galarraga, H.; Warren, R.; Lados, D.; Dehoff, R.R.; Kirka, M.M.; Nandwana, P. Effects of heat treatments on microstructure and properties of Ti-6Al-4V ELI alloy fabricated by electron beam melting (EBM). *Mater. Sci. Eng. A* **2017**, *685*, 417–428. [CrossRef]
13. Benzing, J.; Hrabe, N.; Quinn, T.; White, R.; Rentz, R.; Ahlfors, M. Hot isostatic pressing (HIP) to achieve isotropic microstructure and retain as-built strength in an additive manufacturing titanium alloy (Ti-6Al-4V). *Mater. Lett.* **2019**, *257*, 126690. [CrossRef]

14. Zhai, Y.; Galarraga, H.; Lados, D. Microstructure, static properties, and fatigue crack growth mechanisms in Ti-6Al-4V fabricated by additive manufacturing: LENS and EBM. *Eng. Fail. Anal.* **2016**, *69*, 3–14. [CrossRef]
15. Yu, J.; Rombouts, M.; Maes, G.; Motmans, F. Material Properties of Ti6Al4V Parts Produced by Laser Metal Deposition. *Phys. Procedia* **2012**, *39*, 416–424. [CrossRef]
16. Vastola, G.; Zhang, G.; Pei, Q.-X.; Zhang, Y.-W. Controlling of residual stress in additive manufacturing of Ti6Al4V by finite element modeling. *Addit. Manuf.* **2016**, *12*, 231–239. [CrossRef]
17. Li, P.; Warner, D.; Fatemi, A.; Phan, N. Critical assessment of the fatigue performance of additively manufactured Ti–6Al–4V and perspective for future research. *Int. J. Fatigue* **2016**, *85*, 130–143. [CrossRef]
18. Kasperovich, G.; Hausmann, J. Improvement of fatigue resistance and ductility of TiAl6V4 processed by selective laser melting. *J. Mater. Process. Technol.* **2015**, *220*, 202–214. [CrossRef]
19. Yang, J.; Yu, H.; Yin, J.; Gao, M.; Wang, Z.; Zeng, X. Formation and control of martensite in Ti-6Al-4V alloy produced by selective laser melting. *Mater. Des.* **2016**, *108*, 308–318. [CrossRef]
20. Keist, J.; Palmer, T.A. Role of geometry on properties of additively manufactured Ti-6Al-4V structures fabricated using laser based directed energy deposition. *Mater. Des.* **2016**, *106*, 482–494. [CrossRef]
21. Wang, T.; Zhu, Y.; Zhang, S.; Tang, H.; Wang, H. Grain morphology evolution behavior of titanium alloy components during laser melting deposition additive manufacturing. *J. Alloy. Compd.* **2015**, *632*, 505–513. [CrossRef]
22. Qian, L.; Mei, J.; Liang, J.; Wu, X. Influence of position and laser power on thermal history and microstructure of direct laser fabricated Ti–6Al–4V samples. *Mater. Sci. Technol.* **2005**, *21*, 597–605. [CrossRef]
23. Dinda, G.; Song, L.; Mazumder, J. Fabrication of Ti-6Al-4V Scaffolds by Direct Metal Deposition. *Met. Mater. Trans. A* **2008**, *39*, 2914–2922. [CrossRef]
24. Vrancken, B.; Thijs, L.; Kruth, J.-P.; Van Humbeeck, J. Heat treatment of Ti6Al4V produced by Selective Laser Melting: Microstructure and mechanical properties. *J. Alloys Compd.* **2012**, *541*, 177–185. [CrossRef]
25. Donachie, M.J. *Titanium: A Technical Guide*; ASM International: Materials Park, OH, USA, 2000.
26. Cao, S.; Hu, Q.; Huang, A.; Chen, Z.; Sun, M.; Zhang, J.; Fu, C.; Jia, Q.; Lim, C.V.S.; Boyer, R.R.; et al. Static coarsening behaviour of lamellar microstructure in selective laser melted Ti–6Al–4V. *J. Mater. Sci. Technol.* **2019**, *35*, 1578–1586. [CrossRef]
27. Han, J.; Yang, J.; Yu, H.; Yin, J.; Gao, M.; Wang, Z.; Zeng, X. Microstructure and mechanical property of selective laser melted Ti6Al4V dependence on laser energy density. *Rapid Prototyp. J.* **2017**, *23*, 217–226. [CrossRef]
28. Ungár, T. Microstructural parameters from X-ray diffraction peak broadening. *Scr. Mater.* **2004**, *51*, 777–781. [CrossRef]
29. Wu, M.-W.; Chen, J.-K.; Tsai, M.-K.; Wang, S.-H.; Lai, P.-H. Intensification of preferred orientation in the additive manufactured Ti-6Al-4V alloy after heat treatment. *Mater. Lett.* **2021**, *286*, 129198. [CrossRef]
30. Kelly, S.M.; Kampe, S.L. Microstructural evolution in laser-deposited multilayer Ti-6Al-4V builds: Part I. Microstructural characterization. *Met. Mater. Trans. A* **2004**, *35*, 1861–1867. [CrossRef]
31. Kok, Y.; Tan, X.; Wang, P.; Nai, M.; Loh, N.; Liu, E.; Tor, S.B. Anisotropy and heterogeneity of microstructure and mechanical properties in metal additive manufacturing: A critical review. *Mater. Des.* **2018**, *139*, 565–586. [CrossRef]
32. Gushchina, M.O.; Ivanov, S.Y.; Vildanov, A.M. Effect of Temperature Field on Mechanical Properties of Direct Laser Deposited Ti-6Al-4V Alloy. *IOP Conf. Ser. Mater. Sci. Eng.* **2020**, *969*, 012103. [CrossRef]

Article

Heat Treatment of Corrosion Resistant Steel for Water Propellers Fabricated by Direct Laser Deposition

Ruslan Mendagaliev [1,2], Olga Klimova-Korsmik [1,2,3], Vladimir Promakhov [3,*], Nikita Schulz [3], Alexander Zhukov [3], Viktor Klimenko [3] and Andrey Olisov [3]

1. Institute of Laser and Welding Technologies, St. Petersburg State Marine Technical University, St. Petersburg 190121, Russia; ruslanm888@mail.ru (R.M.); o.klimova@ltc.ru (O.K.-K.)
2. Institute of Metallurgy, Mechanical Engineering and Transport, Peter the Great St. Petersburg Polytechnic University, St. Petersburg 195251, Russia
3. SEC Siberian Center for Industrial Design and Prototyping, National Research Tomsk State University, Tomsk Oblast 634050, Russia; schulznikita97@gmail.com (N.S.); zhuk_77@mail.ru (A.Z.); klimenko@siberia.design (V.K.); science@klimenko.team (A.O.)
* Correspondence: vvpromakhov@mail.ru; Tel.: +7-962-787-21-28

Received: 16 May 2020; Accepted: 15 June 2020; Published: 17 June 2020

Abstract: The urgency of heat treatment of samples of maraging steel obtained by direct laser deposition from steel powder 06Cr15Ni4CuMo is considered. The structural features and properties of 06Cr15Ni4CuMo steel samples after direct laser deposition and heat treatment are studied. The work is devoted to research into the influence of thermal processing on the formation of structure and the mechanical properties of deposit samples. Features of formation of microstructural components by means of optical microscopy are investigated. Tests for tension and impact toughness are conducted. As a result, it was established that the material obtained by the direct laser deposition method in its initial state significantly exceeds the strength characteristics of heat treatment castings of similar chemical composition, but is inferior to it in terms of impact toughness and relative elongation. The increase in relative elongation and impact toughness up to the level of cast material in the deposit samples is achieved at the subsequent heat treatment, which leads to the formation of the structure of tempered martensite and reduction in its content at two-stage tempering in the structure of the metal. The strength of the material is also reduced to the level of cast metal.

Keywords: direct laser deposition (DLD); direct metal deposition; additive manufacturing (AM); corrosion resistant steel; heat treatment (HT); maraging steel; microstructure; mechanical characteristics

---

1. Introduction

Currently, to increase the competitiveness of shipyards for the manufacture of parts of marine engineering, new high-tech technologies are used. Additive manufacturing methods are increasingly being used, including direct laser deposition technology (DLD). In the DLD process it is possible to obtain parts, including from shipbuilding steels used in the Arctic. Iron and its modified alloys are the most important class of metallic materials used in shipbuilding. Different grades of stainless steels can be treated as a part of traditional manufacturing techniques such as casting, machining, powder metallurgy and welding, including any combination of those. [1,2].

As a result of the research on estimation of material characteristics in 1980, this has been developed and mastered in the industry of martensitic–austenitic stainless steel (CA6NM—06Cr15Ni4CuMo) for manufacturing compressors [3], propeller blade [4–7], castings of blades, components of chemical and oil industry and other cast details of responsible purpose, and now for the manufacture of large castings for a propeller blade of blades and hub steel of 06Cr15Ni4CuMo. Currently, it is relevant to obtain parts and blanks for industrial production of this steel by the DLD method.

With the development of modern production technologies, it has become possible to manufacture parts from virtually any metal powder using additive technologies (AM). AM is characterized by quick fabrication and economical spending of expensive materials. Direct laser deposition (DLD) methods are of special interest when large workpieces need to be made using AM. The DLD technology makes it possible to fabricate large parts from stainless and cold-resistant steels [8–12]. The features of the DLD process include high temperature gradients, and repeated fast heating and fast cooling that cause residual strains and form heterogeneities in the microstructure. Microstructural features, such as grain size and morphology (as well as phase transitions), are very sensitive to the dynamic thermal history and they directly influence the microhardness, tear strength and modulus of resilience [13–18].

The mechanical properties of alloys obtained using DLD depend on their structure-phase states. For steels, an important role is played by the content of laminar low-carbon martensite (α') microstructure and the presence of other phases, such as delta ferrite (δ), austenite (γ) and chromium carbides. It is known that the austenite phase is preserved in the tempering process and/or it is restored as a result of the treatment of quenched material. The secondary phase of ferrite in the alloy is δ-ferrite, which forms at very high temperatures during hardening in the course of casting or the DLD process [19].

To achieve high mechanical characteristics for martensite stainless steel, heat treatment (HT) is normally used. However, the microstructure of alloys obtained using DLD is close to a cast structure and is anisotropic. Such materials are characterized by low plasticity and modulus of resilience [20–22]. Here, the material's inner structure is particularly influenced by cyclic heating during DLD due to layer-by-layer metal deposition [23–25]. During the fabrication process, the work piece is tempered. While classical casting and welding envisages a full HT cycle (quenching and tempering), parts fabricated by deposition require the development of ad-hoc HT that is different from the classical one [26,27]. The influence of these effects on the structural characteristics and mechanical properties needs research, specifically the presence of residual austenite γ, non-quenched martensite α', and delta ferrite δ. These factors govern the final properties of the manufactured parts.

The goal of the research work was to grow a plate of 06Cr15Ni4CuMo steel and further reveal the patterns of microstructure formation and mechanical properties after high tempering and determine the maintenance regime providing the required level of ductility of steel not inferior to casting.

2. Materials and Methods

2.1. Materials

We have chosen 06Cr15Ni4CuMo (an analog of CA6NM) for the material. The starting material is 06Cr15Ni4CuMo fraction 45–160 µm Figure 1 is the producer of the "Polema" powder. The chemical composition of the steel is provided in Table 1.

The impact bending tests of the 06Cr15Ni4CuMo steel were conducted on an RKP 450 (Zwick/Roell, Ulm, Germany) unit at −10 °C, with impact energy 150 J and tensile tests were performed on a Z100 (Zwick/Roell, Ulm, Germany) unit at room temperature. Samples made under mechanical tensile testing GOST 1497-84 (RU) and impact testing GOST 6996-66 (RU).

To show the structure, we used chemical etching with Kalling's reagent in a solution (33 mL HCl + 33 mL of ethanol + 33 mL of H_2O + 1.5 g of $CuCl_2$) over 30–60 s.

The deposited cladding layers were visually examined and instrumentally measured; then, they were investigated by optical microscopy on the DMI 500 Leica (Leica Microsystems, Wetzlar, Germany) microscopes using Thixomet (Thixomet, St.-Petersburg, Russia).

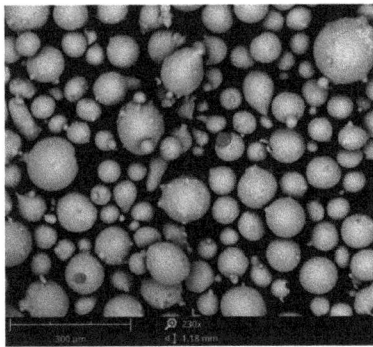 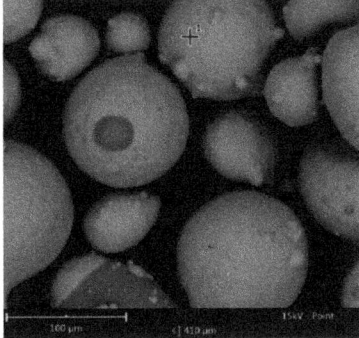

Figure 1. Surface of powder particles 06Cr15Ni4CuMo.

Table 1. The chemical composition of the steel.

Material Grade	Elements Mass Ratio, %									
	C	Si	Mn	Cr	Ni	Mo	S	P	Fe	Cu
06Cr15Ni4CuMo	≤0.06	0.40	0.60–0.90	14.0–15.5	4.0–4.4	0.11–0.28	0.015	0.015	Bal.	1.0–1.5

2.2. Fabrication of Samples Using DLD and Their Heat Treatment

We used the following equipment: a robotic complex based on an LRM-200iD_7L (Fanuc, Oshino, Japan) industrial robot; fiber laser based on a LS-3 Yb (IRE Polus Ltd., Fryazino, Moscow Region, Russia) unit; an FLW D30 (IPG Photonics, Oxford, UK) laser deposition head with a detachable SO12 (Fraunhofer IWS, Aachen, Germany) deposition nozzle and a Twin 10C (Sulzer Metco Inc., New York, NY, USA) powder feeder. A shielding gas atmosphere was used for deposition: an air-proof chamber filled with argon at an excess pressure of 2–3 MPa. In the argon-filled chamber the content of oxygen was not more than 300 ppm. Manufacturing of samples by DLD method was carried out at power P = 2300 W, speed V = 25 mm/s and cross section of depositing bead 2 × 0.8 mm² powder flow rate G = 35 g/min, displacement along the Δx = 1.6 mm, and Δz = 0.7 mm. Five blocks (1 in the initial state and 4 per HT) with LxWxH dimensions 130 × 80 × 16 mm in Figure 2 were deposit simultaneously: a layer was applied alternately to each of the samples, after that the transition to the next layer took place.

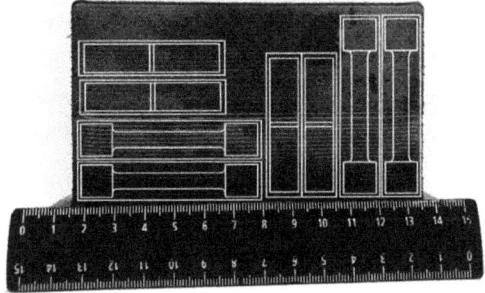

Figure 2. The scheme of cutting samples for mechanical testing.

DLD was carried out in a shielding chamber with controllable atmosphere. High purity argon was used as a transport and protective gas. Overpressure of 2–3 Mbar was maintained in the deposition chamber. The residual oxygen content in the working atmosphere did not exceed 300 ppm. The porosity in the grown plates did not exceed 2% of the total volume of the deposit sample.

The heat treatment was performed in a SNOL 30/1300 muffle furnace without, shielding gas. The heating rate was 200 °C per hour with subsequent exposure and cooling as shown in Figure 3.

Cooling rate of samples after high-temperature tempering was 50 °C per hour and then air cooling was 150 °C.

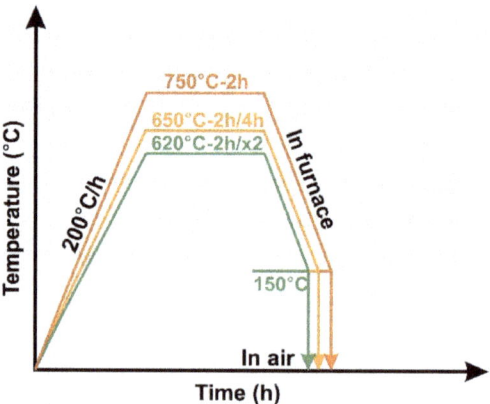

Figure 3. The heat treatment of the samples fabricated by deposition.

3. Results

In the course of parts fabrication from martensite grade steels using DLD, forced process thermal cycling is taking place due to extensive heat deposition. Heat deposition then impacts the presence of retained austenite and δ-ferrite [28] in the structure of the samples fabricated by DLD. Residual austenite can have a negative effect on hardness and toughness.

In the course of tempering during subsequent cooling, metal plasticity deteriorates, and this is due to the formation of secondary martensite as a result of conversion of residual austenite. That is why it makes sense to conduct a second tempering of secondary martensite. This promotes an increase in the metal's relative elongation, creating a finer structure as a result of the decomposition of secondary martensite and the formation of quenched martensite.

Cast metal needs HT for quenching and dual subsequent tempering. During DLD, however, forced thermal cycling is taking place. This strengthens the samples, so we only need to conduct dual tempering to achieve the desired results. Based on the mentioned data from the literature and the results of experimental studies of the steel's characteristics, we have found HT modes for the samples. These modes envisaged high-temperature tempering that would provide the best combination of mechanical properties: high strength, elongation, and impact strength [27].

After DLD the steel has high strength characteristics and low plasticity. To achieve the required mechanical properties for the steel, we have selected several HT modes, as shown in Table 2.

Different δ-morphologies were clearly revealed after etching the fabricated sample, Figure 4a. These morphologies are created by the incomplete growth of Widmanstetten γ-grains during the solid-state δ→γ phase transformation, Figure 4b. Here, incomplete growth results in residual δ stringer inclusions being left at the borders [4]. These inclusions outline the growth front that resembles the initial orientation of the ex- Widmanstett patterns in the final microstructure. During further cooling, γ-austenite converts into α'-martensite, but some amount of δ-delta-ferrite remains in the final microstructure, as shown in Figure 4a. Afterwards, the fabrication δ-delta-ferrite was discovered in the samples, and its content did not exceed 5%.

Table 2. Mechanical properties for heat treatment (HT) modes.

	P (W)	V (mm/s)	Yield Strength, σ_B, (MPa)	Ultimate Strength, $\sigma_{0.2}$, (MPa)	Relative Elongation, δ_s, (%)	Impact Toughness, KV^{-10}, (J)
Technical specifications	N/A	N/A	≥790	≥620	≥19	≥40
DLD	2300	25	1088	792	8	17
			b/T = 750 °C, t = 2 h			
Mode 1	1840	20	1114	798	7.5	16
			c/T = 650 °C, t = 2 h			
Mode 2	1840	20	863	530	15	39
			d/T = 650 °C, t = 4 h			
Mode 3	2300	25	891.7	587.2	12	29
			e/T = 620 °C, t = 2 h			
Mode 4	2300	20	816.4	698.3	16	42
			f/T = 620 °C, t = 2 h/×2 the cycle is repeated twice			
Mode 5	2300	20	804.4	666.8	19	42

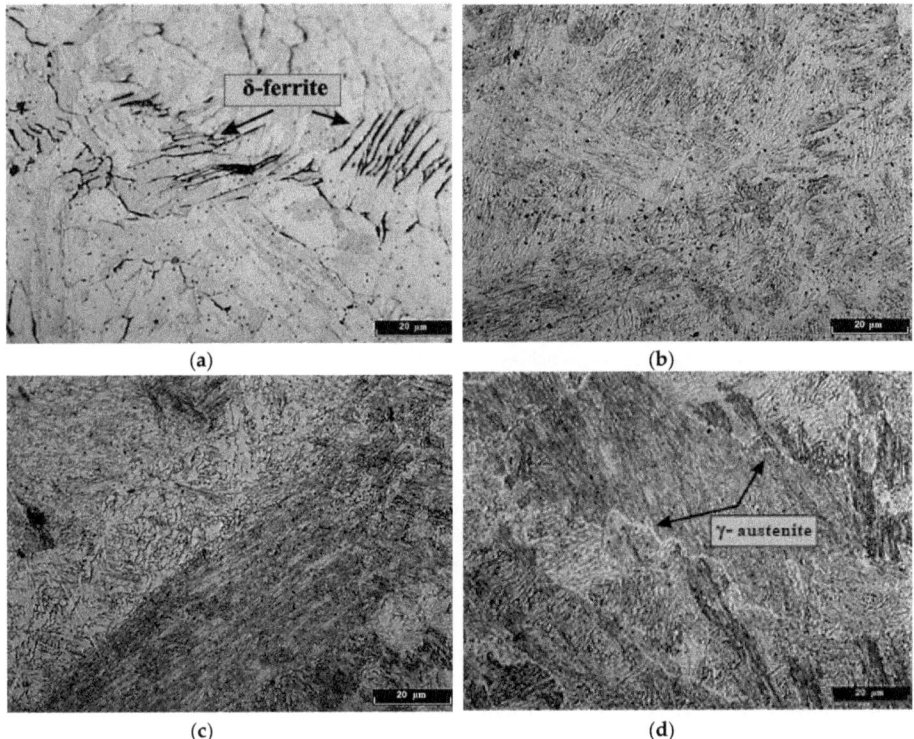

(a) (b)

(c) (d)

Figure 4. Cont.

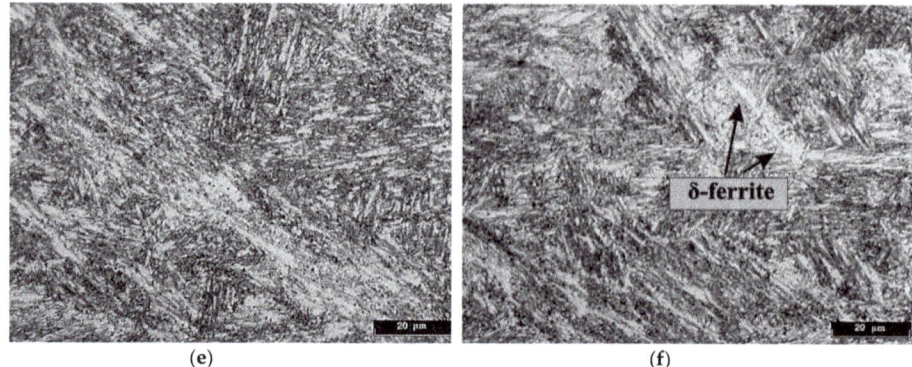

(e) (f)

Figure 4. The microstructure of 06Cr15Ni4CuMo (**a**) DLD; (**b**) mode 1 (T = 750 °C, t = 2 h); (**c**) mode 2 (T = 650 °C, t = 2 h); (**d**) mode 3 (T = 650 °C, t = 4 h); (**e**) mode 4 (T = 620 °C, t = 2 h); (**f**) mode 5 (T = 620 °C, t = 2 h/x2).

At such high temperatures, δ-grains are growing rapidly during heating. Then, during the cooling process, they convert into a γ-phase with subsequent transition into a α' structure. In Figure 4b, after a single iteration of high-temperature tempering, some amount of γ-austenite (as well as M_7C_3 chromium carbides) is still observed.

Figure 4c,d shows some amount of non-converted α'-martensite that is an unstable microstructure in the α matrix. It promotes the formation of carbides, interlayer boundaries at large angles and a γ-phase.

Figure 4e,f includes a structure after high-temperature tempering that is represented by quenched lath α'-martensite with fine particles of residual austenite. The particles are located between martensite laths and at the boundaries of martensite batches with large inclusions of δ-ferrite in the matrix basis.

The best mechanical properties were achieved at dual tempering at T = 620 °C, t = 2 h/x2 the cycle is repeated twice Figure 4. This mode complies with the technical specifications for this steel grade while having slightly lower plasticity characteristics.

Figure 5 shows the distribution of microhardness single weld bed and their arrangement for different HT modes and a thermokinetic diagram for Fe-Cr-Ni.

The average microhardness in the fabricated sample tempered at 750 °C, Figure 5b, was 355 HV. Tempering at 750 °C was chosen because of the phase transition into the γ-austenity area that preceded the dissolution of existing chromium carbides in the fabricated sample (i.e., $M_{23}C_6/M_7C_3$). Thus, it increased the concentration of carbon in the martensite matrix at room temperature.

With temperature reduction to 650 °C it was possible to decrease the hardness to 280 HV with an exposure time of 2 h, and an exposure time of 4 h was required to decrease it to 301 HV. At high-temperature tempering at T = 620 °C with an exposure time of 2 h, the microhardness was 260 HV, and it was 273 HV after the second tempering.

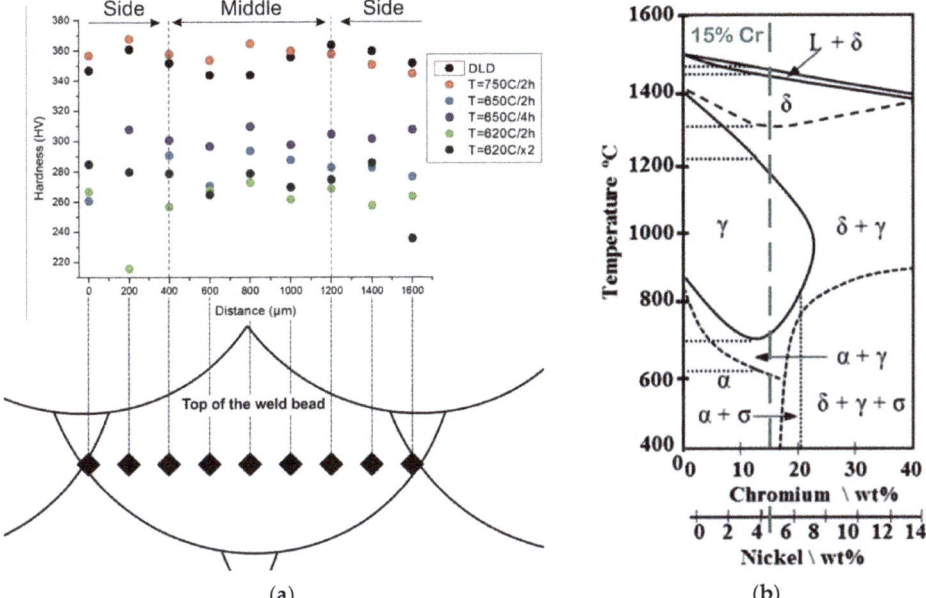

Figure 5. (a) microhardness; (b) the Fe–Cr–Ni equilibrium phase diagram [29].

4. Discussion

It is established that the DLD process is the fastest and most convenient for creating parts. After the process, it is necessary to produce high tempering to achieve all the necessary mechanical properties. Samples obtained by the DLD method are not inferior in characteristics to casting, and in some cases are most in demand.

Thus, we can conclude that the propeller with an optimized structure has reliability characteristics close to the original solid version. We showed the manufacturing of hub and blades via DLD and built-up propeller before and after CNC-machining and manual polishing. After all producing stages, the propeller was weighted. The weight test showed that the mass is finished product is 105 kg. This is 20% less than the original cast design. A more detailed description of design analysis, optimization procedure and production process is presented in [30].

5. Conclusions

The DLD process of 06Cr15Ni4CuMo steel achieves a high strength due to the forced thermal cycling process at low impact toughness and relative elongation. In order to eliminate the imbalance of the complex of mechanical properties, HT is proposed.

According to the results of a comparison of mechanical properties, it was established that the lowest structural matrices are tempered by fine martensite with a low level of residual austenite and δ-ferrite. Based on the analysis of the relationship between HT, mechanical properties and 06Cr15Ni4CuMo steel structure, the most suitable heat treatment mode for deposit samples was established, consisting of a double HT at T = 620 °C, t = 2 h/x2; the cycle is repeated twice, in which a finely dispersed structure of tempered rack martensite is formed, providing a set of properties equal to the material obtained by casting.

Author Contributions: Conceptualization, R.M. and V.P.; methodology, V.P., V.K. and O.K.-K.; validation, A.Z. and N.S.; resources, V.K. and V.P.; writing—original draft preparation, R.M., A.O., O.K.-K., and V.P.; writing—review and editing, N.S. and R.M.; visualization, A.Z.; supervision, R.M. and A.O. All authors have read and agreed to the published version of the manuscript.

Funding: This research was funded by RFBR and Tomsk Region, project number 19-48-703019, was supported by The Tomsk State University competitiveness improvement program (grant No. 8.2.04.2020), grant SP-724.2019.1.

Conflicts of Interest: The authors declare no conflict of interest. The funders had no role in the design of the study; in the collection, analyses, or interpretation of data; in the writing of the manuscript, or in the decision to publish the results.

References

1. Khodabakhshi, F.; Farshidianfar, M.; Gerlich, A.; Nosko, M.; Trembošová, V.; Khajepour, A. Microstructure, strain-rate sensitivity, work hardening, and fracture behavior of laser additive manufactured austenitic and martensitic stainless steel structures. *Mater. Sci. Eng. A* **2019**, *756*, 545. [CrossRef]
2. Promakhov, V.; Zhukov, A.; Ziatdinov, M.; Zhukov, I.; Schulz, N.; Kovalchuk, S.; Dubkova, Y.; Korsmik, R.; Klimova-Korsmik, O.; Turichin, G.; et al. Inconel 625/TiB$_2$ Metal Matrix Composites by Direct Laser Deposition. *Metals* **2019**, *9*, 141. [CrossRef]
3. Winck, L.B.; Ferreira, J.; Araújo, J.; Manfrinato, M.D.; Silva, C. Surface nitriding influence on the fatigue life behavior of ASTM A743 steel type CA6NM. *Surf. Coat. Technol.* **2013**, *232*, 844. [CrossRef]
4. Wen, P.; Cai, Z.; Feng, Z.; Wang, G. Microstructure and mechanical properties of hot wire laser clad layers for repairing precipitation hardening martensitic stainless steel. *Opt. Laser Technol.* **2015**, *75*, 207. [CrossRef]
5. Promakhov, V.; Khmeleva, M.; Zhukov, I.A.; Platov, V.V.; Khrustalyov, A.; Vorozhtsov, A.B. Influence of Vibration Treatment and Modification of A356 Aluminum Alloy on Its Structure and Mechanical Properties. *Metals* **2019**, *9*, 87. [CrossRef]
6. Mirakhorli, F.; Cao, X.; Pham, X.-T.; Wanjara, P.; Fihey, J. Phase structures and morphologies of tempered CA6NM stainless steel welded by hybrid laser-arc process. *Mater. Charact.* **2017**, *123*, 264. [CrossRef]
7. Zhukov, I.A.; Kozulin, A.A.; Khrustalyov, A.; Matveev, A.E.; Platov, V.V.; Vorozhtsov, A.B.; Zhukova, T.V.; Promakhov, V. The Impact of Particle Reinforcement with Al$_2$O$_3$, TiB$_2$, and TiC and Severe Plastic Deformation Treatment on the Combination of Strength and Electrical Conductivity of Pure Aluminum. *Metals* **2019**, *9*, 65. [CrossRef]
8. Turichin, G.; Zemlyakov, E.; Babkin, K.; Ivanov, S.; Vildanov, A. Analysis of distortion during laser metal deposition of large parts. *Procedia CIRP* **2017**, *74*, 154. [CrossRef]
9. Turichin, G.; Zemlyakov, E.; Kuznetsov, M.; Babkin, K.; Kurakin, A.; Vildanov, A. Direct laser deposition with transversal oscillating of laser radiation. In Proceedings of the International Conference Laser Optics, St. Petersburg, Russia, 4–8 June 2018; p. 124. [CrossRef]
10. Turichin, G.; Kuznetsov, M.; Tsibulskiy, I.; Firsova, A. Hybrid Laser-Arc Welding of the High-Strength Shipbuilding Steels: Equipment and Technology. *Phys. Procedia* **2017**, *89*, 156. [CrossRef]
11. Rahman, N.U.; Capuano, L.; Cabeza, S.; Feinaeugle, M.; Garcia-Junceda, A.; De Rooij, M.; Matthews, D.; Walmag, G.; Gibson, I.; Römer, G. Directed energy deposition and characterization of high-carbon high speed steels. *Addit. Manuf.* **2019**, *30*. [CrossRef]
12. Zhao, X.; Dong, S.; Yan, S.; Liu, X.; Liu, Y.; Xia, D.; Lv, Y.; He, P.; Xu, B.; Han, H. The effect of different scanning strategies on microstructural evolution to 24CrNiMo alloy steel during direct laser deposition. *Mater. Sci. Eng. A* **2019**. [CrossRef]
13. Turichin, G.; Klimova, O.; Zemlyakov, E.; Babkin, K.; Kolodyazhnyy, D.; Shamray, F.; Travyanov, A.; Petrovskiy, P. Technological aspects of high speed direct laser deposition based on heterophase powder metallurgy. *Phys. Procedia* **2015**, *78*, 397. [CrossRef]
14. Turichin, G.A.; Zemlyakov, E.V.; Klimova, O.G.; Babkin, K.D. Hydrodynamic instability in high-speed direct laser deposition for additive manufacturing. *Phys. Procedia* **2016**, *83*, 674. [CrossRef]
15. Glukhov, V.; Turichin, G.; Klimova-Korsmik, O.; Zemlyakov, E.; Babkin, K. Quality Management of Metal Products Prepared by High-Speed Direct Laser Deposition Technology. *Key Eng. Mater.* **2016**, *684*, 461. [CrossRef]
16. Turichin, G.; Kuznetsov, M.; Pozdnyakov, A.; Gook, S.; Gumenyuk, A.; Rethmeier, M. Influence of heat input and preheating on the cooling rate, microstructure and mechanical properties at the hybrid laser-arc welding of API 5L X80 steel. *Procedia CIRP* **2018**, *74*, 748. [CrossRef]
17. Turichin, G.; Zemlyakov, E.; Babkin, K.; Ivanov, S.; Vildanov, A. Laser metal deposition of Ti-6Al-4V alloy with beam oscillation. *Procedia CIRP* **2018**, *74*, 184. [CrossRef]

18. Klimova-Korsmik, O.; Turichin, G.; Zemlyakov, E.; Babkin, K.; Petrovsky, P.; Travyanov, A. Structure formation in Ni superalloys during high-speed direct laser deposition. *Mater. Sci. Forum* **2017**, *879*, 978. [CrossRef]
19. Trudel, A.; Lévesque, M.; Brochu, M. Microstructural effects on the fatigue crack growth resistance of a stainless steel CA6NM weld. *Eng. Fract. Mech.* **2014**, *115*, 60. [CrossRef]
20. Fang, J.; Dong, S.; Li, S.; Wang, Y.; Xu, B.; Li, J.; Liu, B.; Jiang, Y. Direct laser deposition as repair technology for a low transformation temperature alloy: Microstructure, residual stress, and properties. *Mater. Sci. Eng. A* **2019**, *748*, 119. [CrossRef]
21. Guan, T.; Chen, S.; Chen, X.; Liang, J.; Liu, C.; Wang, M. Effect of laser incident energy on microstructures and mechanical properties of 12CrNi2Y alloy steel by direct laser deposition. *J. Mater. Sci. Technol.* **2019**, *35*, 395. [CrossRef]
22. Wei, S.; Wang, G.; Yu, J.; Rong, Y. Competitive failure analysis on tensile fracture of laser-deposited material for martensitic stainless steel. *Mater. Des.* **2017**, *118*, 1. [CrossRef]
23. Sarafan, S.; Wanjara, P.; Champliaud, H.; Thibault, D. Characteristics of an autogenous single pass electron beam weld in thick gage CA6NM steel. *Int. J. Adv. Manuf. Technol.* **2015**, *78*, 1523. [CrossRef]
24. Barros, R.; Silva, F.J.G.; Gouveia, R.M.; Saboori, A.; Marchese, G.; Biamino, S.; Salmi, A.; Atzeni, E. Laser Powder Bed Fusion of Inconel 718: Residual Stress Analysis Before and After Heat Treatment. *Materials* **2020**, *13*, 2248. [CrossRef]
25. Liu, Y.; Li, A.; Cheng, X.; Zhang, S.; Wang, H. Effects of heat treatment on microstructure and tensile properties of laser melting deposited AISI 431 martensitic stainless steel. *Mater. Sci. Eng. A* **2016**, *666*, 27. [CrossRef]
26. Gouveia, R.M.; Silva, F.J.G.; Atzeni, E.; Sormaz, D.; Alves, J.L.; Pereira, A.B. Effect of Scan Strategies and Use of Support Structures on Surface Quality and Hardness of L-PBF AlSi10Mg Parts. *Materials* **2020**, *13*, 2248. [CrossRef]
27. Rashid, R.A.R.; Nazari, K.; Barr, C.; Palanisamy, S.; Orchowski, N.; Matthews, N.; Dargusch, M. Effect of laser reheat post-treatment on the microstructural characteristics of laser-cladded ultra-high strength steel. *Surf. Coat. Technol.* **2019**, *372*, 93. [CrossRef]
28. Tsukanov, V.V.; Tsyganko, L.K.; Petrov, S.N.; Shandyba, G.A.; Ziza, A.I. Structural transformations during heat treatment of cast corrosion-resistant martensitic steel. *Metallobrabotka* **2016**, *3*, 42.
29. Folkhard, E.; Rabensteiner, G.; Perteneder, E. *Welding Metallurgy of Stainless Steels*; Springer: Vienna, Austria, 1988; p. 98.
30. Korsmik, R.; Rodionov, A.; Korshunov, V.; Ponomarev, D.; Prosychev, I.; Promakhov, V. Topological optimization and manufacturing of vessel propeller via LMD-method. *Mater. Today Proc.* **2020**. [CrossRef]

 © 2020 by the authors. Licensee MDPI, Basel, Switzerland. This article is an open access article distributed under the terms and conditions of the Creative Commons Attribution (CC BY) license (http://creativecommons.org/licenses/by/4.0/).

Article

Effect of Elevated Temperatures on the Mechanical Properties of a Direct Laser Deposited Ti-6Al-4V

Sergei Ivanov [1,*], Marina Gushchina [1], Antoni Artinov [2], Maxim Khomutov [3] and Evgenii Zemlyakov [1]

Citation: Ivanov, S.; Gushchina, M.; Artinov, A.; Khomutov, M.; Zemlyakov, E. Effect of Elevated Temperatures on the Mechanical Properties of a Direct Laser Deposited Ti-6Al-4V. *Materials* 2021, *14*, 6432. https://doi.org/10.3390/ma14216432

Academic Editor: Amir Mostafaei

Received: 28 September 2021
Accepted: 22 October 2021
Published: 27 October 2021

Publisher's Note: MDPI stays neutral with regard to jurisdictional claims in published maps and institutional affiliations.

Copyright: © 2021 by the authors. Licensee MDPI, Basel, Switzerland. This article is an open access article distributed under the terms and conditions of the Creative Commons Attribution (CC BY) license (https://creativecommons.org/licenses/by/4.0/).

[1] World-Class Research Center "Advanced Digital Technologies", St. Petersburg State Marine Technical University, Lotsmanskaya 3, 190121 St. Petersburg, Russia; gushcina_mo@corp.smtu.ru (M.G.); e.zemlyakov@ltc.ru (E.Z.)
[2] Bundesanstalt für Materialforschung und -Prüfung (BAM), Unter den Eichen 87, 12205 Berlin, Germany; Antoni.Artinov@bam.de
[3] Department of Physical Metallurgy of Non-Ferrous Metals, National University of Science and Technology "MISiS", Leninskiy Prospekt 4, 119049 Moscow, Russia; khomutov@misis.ru
* Correspondence: Sergei.Yu.Ivanov@gmail.com

Abstract: In the present work, the mechanical properties of the DLD-processed Ti-6Al-4V alloy were obtained by tensile tests performed at different temperatures, ranging from 20 °C to 800 °C. Thereby, the process conditions were close to the conditions used to produce large-sized structures using the DLD method, resulting in specimens having the same initial martensitic microstructure. According to the obtained stress curves, the yield strength decreases gradually by 40% when the temperature is increased to 500 °C. Similar behavior is observed for the tensile strength. However, further heating above 500 °C leads to a significant increase in the softening rate. It was found that the DLD-processed Ti-6Al-4V alloy had a Young's modulus with higher thermal stability than conventionally processed alloys. At 500 °C, the Young's modulus of the DLD alloy was 46% higher than that of the wrought alloy. The influence of the thermal history on the stress relaxation for the cases where 500 °C and 700 °C were the maximum temperatures was studied. It was revealed that stress relaxation processes are decisive for the formation of residual stresses at temperatures above 700 °C, which is especially important for small-sized parts produced by the DLD method. The coefficient of thermal expansion was investigated up to 1050 °C.

Keywords: direct laser deposition; Ti-6Al-4V; mechanical properties; microstructure; stress relaxation; elevated temperatures

1. Introduction

Direct Laser Deposition (DLD) is one of the most widely utilized additive manufacturing (AM) technologies for the production of Ti-6Al-4V alloy parts. The uneven local heating of the buildup during DLD leads to significant stresses and distortion that affect the service properties and the shape of the final parts [1–3]. In the last, decade numerous models have been proposed for the simulation of these phenomena [4–6]. Thereby, the temperature dependence of the mechanical properties is a major factor influencing the accuracy of the simulation other than the assumptions of the mathematical model. Nowadays, it is a common practice to use the material properties of wrought Ti-6Al-4V alloys for the simulation of the DLD process. Mukherjee et al. [7] conducted a thermo-mechanical simulation of a DLD-processed Ti-6Al-4V-alloy by considering the material properties as being temperature dependent in the temperature range between 20 °C and 1600 °C. In their study, a combination of material properties derived from the fully lamellar wrought alloy [8] and the Ti-6Al-4V metal matrix composite [9] was used rather than the properties of the DLD-processed Ti-6Al-4V alloy. Denlinger and Michaleris [10] found a significant difference between the numerically predicted and experimentally measured distortion when using the mechanical properties of the wrought Ti-6Al-4V. Lu et al. [11]

revealed significant scattering of the available data on the mechanical properties of the wrought Ti-6Al-4V alloy at elevated temperatures. Furthermore, a sensitivity analysis of the mechanical properties of Ti-6Al-4V showed that the distortion and the residual stresses strongly depend on the thermal expansion coefficient and less on the Young's modulus and the elastic limit. They concluded that for the numerical analysis of the AM process, it is mandatory to use material properties that are specific to the particular manufacturing process. The present study is intended to fill the gap in the lack of data on the temperature dependence of the mechanical properties of DLD-processed Ti-6Al-4V alloys.

The poor mechanical properties of a commercially pure titanium hinder its application as a structural material. However, critical parts that require high strength and ductility, corrosion resistance in aggressive environments, heat resistance, etc., are made of titanium alloys. The Ti-6Al-4V alloy used in this study is a two-phase ($\alpha + \beta$) alloy. Recognized as the most popular titanium alloy, Ti-6Al-4V occupies almost a half of the market share of titanium products used in the world today. The proportion of Al and V results in the material having attractive mechanical properties. Ti-6Al-4V contains 6 wt% Al, which stabilizes the α-phase of the hexagonal close-packed structure and 4 wt% V, which stabilizes the β-phase of the body-centered cubic structure. The two phases have different properties due to their structures, with α exhibiting greater strength yet lower ductility and formability [12]. The aluminum in the alloy increases the strength and heat-resistant properties, whereby vanadium increases not only the strength properties but also increases the ductility. It is well-known that two-phase titanium alloys have a lower sensitivity to hydrogen, e.g., hydrogen-induced cold cracking, compared to pseudo-α-alloys. Furthermore, they have good manufacturability and a relatively low tendency to undergo salt corrosion [13–15]. Titanium alloys have a good castability due to the short solidification interval of less than 50–70 °C [15,16]. Hereby, the chemical composition of the Ti-6Al-4V alloy utilized in casting does not differ from that of the wrought alloy [17]. The Ti-6Al-4V alloy utilized in additive manufacturing only has a slight difference in the content of carbon impurities from the wrought alloy [18].

It is a well-known fact that the mechanical and service properties of the alloy are determined by the microstructure. The high ductility and cyclic strength correspond to an equiaxed fine grain microstructure. On the other hand, the lamellar microstructure has a high fracture toughness and greater crack propagation resistance. Therefore, it can be said that the bimodal (duplex) microstructure offers an optimal combination of the mechanical properties of the wrought Ti-6Al-4V alloy. The control and optimization of the morphology of the α phase is one of the important issues in terms of the use of the alloy. Thermomechanical processing is a very useful method for improving the microstructure, e.g., controlling the size and the aspect ratio of the α lamellar phase, optimizing the phase ratio of the α to β phases, and controlling the morphology of the β phase [19,20]. The microstructure of the Ti-6Al-4V alloy obtained by direct laser deposition (DLD) depends strongly on the heat input and the inter-pass temperature [1,21], the variation of which results in a wide range of obtained mechanical properties. It is worth noting that the ductility of a DLD-processed alloy can fall to near zero, whereas the strength properties remain comparable to those of the wrought alloy [22].

The effect of the microstructure on the short-term strength of the Ti-6Al-4V alloy at elevated temperatures is similar to its effect on its strength at room temperature. The best combination of ductility, fracture toughness, heat resistance, and endurance is found in alloys with a 70–80% lamellar microstructure [15]. In [23], it was found that alloys with basket-weave microstructures exhibit the most obvious work hardening behavior and the highest strength during hot tensile deformation by temperatures of about 800 °C. The best ductility corresponds to alloys with an equiaxed microstructure. According to [24], the alloy with initial equiaxed microstructures also showed the highest ductility during tensile testing in the temperature range of 20–600 °C, while the material with initial full martensite microstructure showed better thermal strength. A critical analysis of the literature showed a significant spread in the experimental data of the short-term strength of the wrought alloy at elevated temperatures, as can be seen from Figure 1 [7,10,25–33]. This can be explained by the variation in the test conditions, the initial microstructure of the specimens, and the loading parameters.

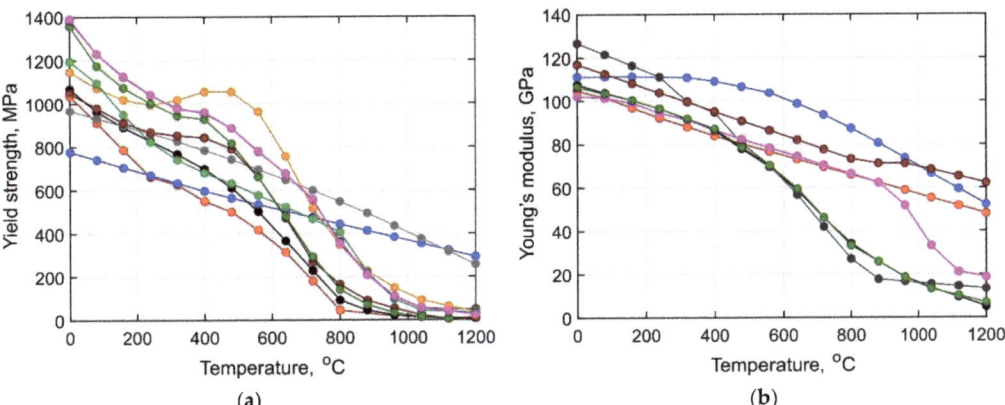

Figure 1. Temperature dependence of the (**a**) yield stress and (**b**) Young's modulus according to [7,10,25–33].

It should be noted that there is practically no data available on the properties of additively manufactured Ti-6Al-4V alloy at elevated temperatures. In [34], an electron beam melting (EBM)-processed material showed a lower flow stress than the wrought alloy during a compression test in the temperature range of 1000–1200 °C. This can be attributed to the larger prior β-grain size and thickness of the α-plates in the EBM-processed alloy. In [35], the flow stress curves of the selective laser melting (SLM)- and direct energy deposition (DED)-processed and wrought alloys were compared in a compression test for temperatures ranging from 850 °C to 1100 °C. It was found that the presence of a percolating β-phase during the decomposition of martensite seems to be the reason for the reduced flow stress of the additively manufactured material compared to conventional wrought material with a lamellar microstructure. DED and SLM materials show a faster transformation to a globular microstructure compared to conventional wrought material. The temperature dependence of the tensile strength of the SLM-processed Ti-6Al-4V alloy in the temperature range between 20 °C and 550 °C was studied in [36]. SLM-processed Ti-6Al-4V alloy showed excellent ultimate tensile strength below 500 °C, which was 100 MPa higher than a solution-treated and aged Ti-6Al-4V alloy and 300 MPa higher than an annealed Ti-6Al-4V alloy. The effect of the strain rate and temperature on the mechanical properties of the DLD-processed alloy with a Widmanstätten microstructure was studied in [37], using a compression and tension test. However, the study lacks a description of the thermal history during the fabrication of the specimens. The presence of defects such as pores and a lack of fusion had a significant impact on the obtained results. The study presented in [38] compares the mechanical properties of a wrought alloy and a SLM-processed alloy in a compression test utilizing strain rates of 0.001–1 s^{-1} in the temperature range of 20–1000 °C. Both publications revealed that the anisotropy of the mechanical properties of the Ti-6Al-4V alloy obtained by various AM methods is insignificant.

The review presented above shows that most of the available material data for additively manufactured Ti-6Al-4V alloy are limited to the study of flow stresses at different strain rates in compression tests above 700 °C. These data are important for determining the parameters of the hot forging or stamping processes but are insufficient for the numerical simulation of stresses and distortion induced by the DLD. It should be noted that to the best of the authors knowledge, none of the publications contain data on the temperature dependence of the Young's modulus of the Ti-6Al-4V alloy obtained with AM methods.

Moreover, the majority of studies are devoted to the investigation of the material properties of SLM-processed alloys, which differ from those of the DLD-processed alloys.

In the present paper, the mechanical properties of a DLD-processed Ti-6Al-4V alloy were obtained through a tensile test performed for different temperatures ranging from 20 °C to 800 °C. The conditions used to obtain the test specimens were close to the conditions used in the manufacturing of large-sized structures by the DLD method. The influence of the thermal history on the stress relaxation for the case of 500 °C and 700 °C maximum temperatures was revealed. In addition, the temperature dependence of the coefficient of thermal expansion was obtained. The influence of the initial microstructure of the samples on the deformation and fractures at elevated temperatures was as well analyzed. An approximation of the measured tensile curves for given temperatures using a proposed fitting function was obtained and used to describe the hardening behavior during plastic deformation.

2. Materials and Methods

2.1. Specimens

Almost all of data published on the mechanical properties of the DLD-processed Ti-6Al-4V alloy refer to samples obtained without or with very short dwell time between the deposited layers, leading to a significant overheating of the buildup. Therefore, in the present study, the tensile samples were machined from buildups with an inter-pass temperature in the range of 60–80 °C, which is typical for large-sized components. Note that the interpass temperature was controlled with type K thermocouples with a diameter of 0.5 mm. The DLD process parameters were as follows: a beam power of 1900 W; a beam diameter of 2.5 mm; a process speed of 20 mm s^{-1}; a powder flow rate of 10.5 g min^{-1}; and a gas flow rate of 25 L min^{-1}. The specimens were made using an in-house robotic DLD machine that was developed at the St. Petersburg State Marine Technical University in St. Petersburg, Russia. The machine included a Fanuc 6500 5-axis industrial robot, a rotary table, and a processing head with a discrete coaxial powder feed. To prevent the oxidation of the specimens during the buildup, the sealed chamber of the machine was filled with argon. Hereby, the residual oxygen content in the chamber did not exceed 100 ppm. In total, 160 layers were deposited, having 12 mm in width, 0.8 mm in height, and 140 mm in length, and each layer consisted of seven passes. Spherical Ti-6Al-4V powder with a diameter of 45–90 µm, which was produced by a plasma rotating electrode method, was used for the buildups. The size distribution of the powder particles was unimodal with no visible non-metallic inclusions on the surface, as shown in Figure 2. The chemical composition was in accordance with the standard ASTM F136-02a [39].

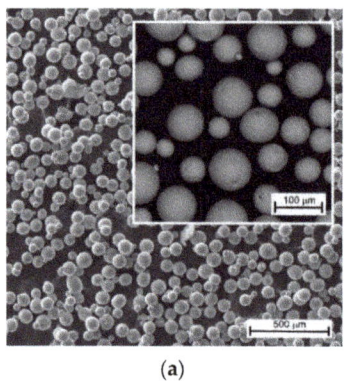
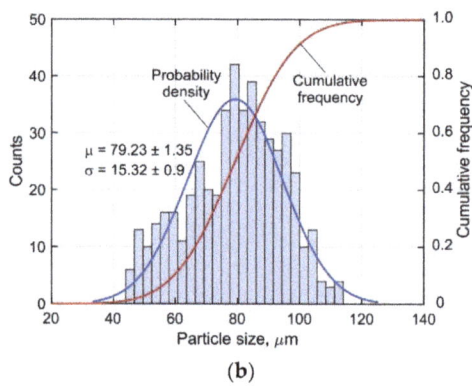

(a)   (b)

Figure 2. (a) Scanning electron micrograph of Ti-6Al-4V powder and (b) size distribution of the powder particles.

2.2. Optical and Scanning Electron Microscopy

Optical metallography of etched microsamples was conducted using a Leica DMI8A microscope with a magnification of up to 1000 times. For the etching, Kroll's reagent (1 mL HF + 2 mL HNO$_3$ + 47 mL H$_2$O) was used [40]. All metallographic cross-sections were taken from the middle of the buildup. The Vickers hardness was measured according to the ISO 6507 standard on an FM-310 hardness tester (Future Tech, Tokyo, Japan) with a load of 3 N. To determine the chemical composition and to analyze the fracture surface of the specimens after testing, a Tescan Mira3 scanning electron microscope (TESCAN, Brno, Czech Republic) with an Oxford AZtec console was used (Oxford Instruments NanoAnalysis, Wycombe, UK).

2.3. Tensile Tests at Elevated Temperatures

The mechanical properties of the DLD-processed Ti-6Al-4V alloy were obtained using a Gleeble 3800 metallurgical simulation system at the National University of Science and Technology MISiS in Moscow, Russia. The setup allows sequential tensile-compression deformation with a force of up to 10 t and simultaneous heating of the sample by direct electric current transmission to be performed. Depending on the specimen configuration and size, the heating and the cooling rate can reach up to 10,000 °C s^{-1} and 3000 °C s^{-1}, respectively. The temperature field was controlled by the contact method using a type K thermocouple with a diameter of 0.25 mm that was fixed to the surface of the sample by discharge spot welding. A schematic of the specimen used for the uniaxial tension tests is shown in Figure 3. Thereby a heating rate of 10 °C s^{-1} and a strain rate of 3 mm min^{-1} were used. An externally mounted sensor was used for the precision recording of the transverse strain. The transverse strain was measured with a 500 Hz sampling rate in the central section of the specimen. The noise in the experimental data, which is shown in Figure 3b, significantly hampered their processing. Therefore, a robust discrete cosine transform (DCT) filter, which was implemented in the commercial software Matlab, was used to process the data [41,42].

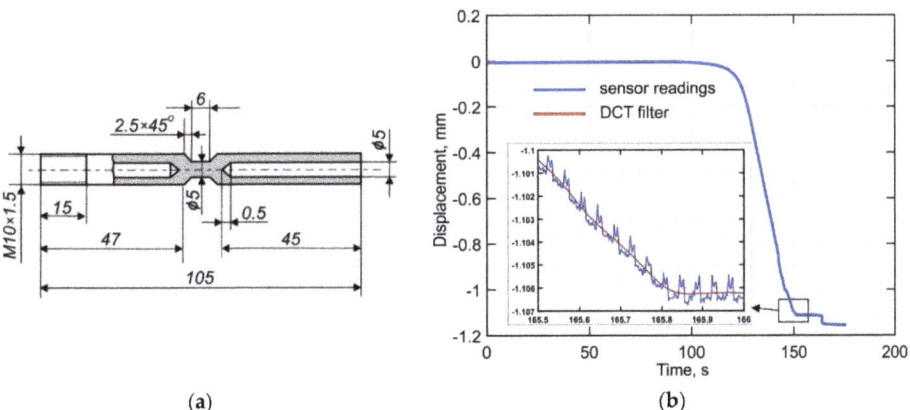

Figure 3. (a) A schematic of the Gleeble 3800 test specimen and (b) an example of experimental data processing using the DCT filter.

An approximation of the measured tensile curves for given temperatures was performed to describe hardening behavior during plastic deformation. The following fitting function was proposed:

$$\sigma(\varepsilon) = \frac{p_1 \cdot \varepsilon^2 + p_2 \cdot \varepsilon + p_3}{\varepsilon + p_3} \cdot \sigma_{0.2}, \tag{1}$$

where σ is the stress, ε is the strain, $\sigma_{0.2}$ is the yield strength, and p_1, p_2, p_3 are the fitting coefficients.

Note that a zero strain in Equation (1) corresponds to a stress level that is equal to the yield strength of the material. The fitting coefficients in Equation (1) were determined by the nonlinear least squares method [43]. Figure 4 shows an example of fitted experimental data. Note that for clarity, only a small part of the recorded experimental points is shown. It can be seen that the proposed fitting function agrees well with the experimental data. However, it should be noted that the fitting of the engineering stress–strain curves was conducted using the data included between the strain corresponding to the yield strength and the strain corresponding to the tensile strength.

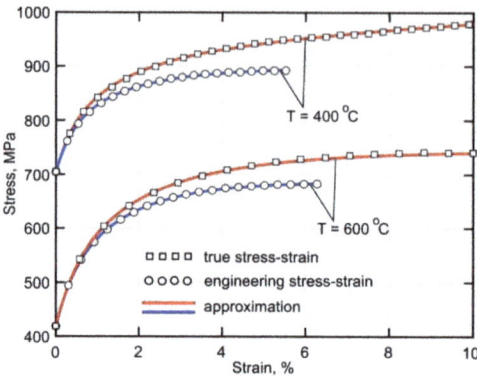

Figure 4. An example of the measured stress–strain curves and their approximation.

2.4. Thermal Expansion Tests

A DIL 805 A/D quenching dilatometer test machine was used to determine the temperature dependent coefficient of thermal expansion (CTE). A cylindrical specimen with a 4 mm diameter and 10 mm length was inductively heated up to 1050 °C at a rate of 3 °C s^{-1}. The tests were conducted in vacuum to prevent oxidation. After holding the sample at the maximum temperature for 20 min, the specimen was cooled at a rate of 0.94 °C s^{-1} by blowing it with helium. An instantaneous α and secant $\bar{\alpha}$ coefficient of thermal expansion were determined, according to Figure 5. Therefore, the following expressions were used:

- for instantaneous CTE:

$$\alpha = \frac{\Delta L_2 - \Delta L_1}{L_o} \cdot \frac{1}{T_2 - T_1}; \qquad (2)$$

- for secant CTE:

$$\bar{\alpha} = \frac{\Delta L(T)}{L_o} \cdot \frac{1}{(T - T_o)}. \qquad (3)$$

The instantaneous CTE was determined by the first derivative of the experimental thermal strain curve with respect to the temperature. Note that an irregular experimental curve can cause significant high-frequency fluctuations of the calculated derivative values. Thus, to obtain a smooth curve of the instantaneous CTE, an experimental thermal strain curve was approximated by piecewise polynomials of 9th degree.

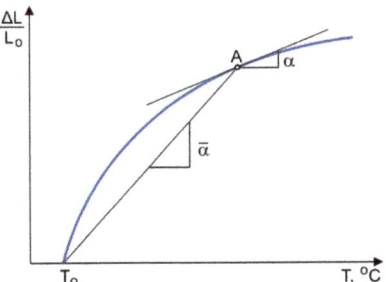

Figure 5. Determination of the coefficients of thermal expansion at point A.

2.5. Stress Relaxation Tests

The visco-plastic behavior of DLD-processed Ti-6Al-4V alloy was measured using a uniaxial tensile test utilized in the Gleeble 3800 machine according to the method described in [44]. Each specimen was first heated to 500 °C or 700 °C and subsequently tensioned to a specified strain value. All samples were tensioned with a strain rate of 3 mm min^{-1} to a total strain of 2%, which is equal to a stress level of about 88% of the yield strength of the alloy at the corresponding temperature. In the next step, the applied strain was held constant, and the stress relaxation was measured as a function of time.

3. Results and Discussion

3.1. Microstructure of the DLD-Processed Ti-6Al-4V Alloy

The following factors affect the microstructure of the Ti-6Al-4V buildup: (1) a high crystallization rate of the deposited metal due to low interpass temperatures [45,46]; (2) multiple short-term irregular reheating phases from subsequent passes [47,48]; (3) epitaxial crystal growth [49,50]. The microstructure consisted of a lamellar α'-phase, as shown in Figure 6, and a small amount of residual β-phase in the form of thin interlayers [51–53]. The residual β-phase cannot be detected by optical or scanning electron microscopy due to its minor content. The presence of an α'-phase leads to an increase in the strength and a decrease in the ductility of the material. The nucleation of the α-phase initiates at the boundaries of the β-grains. The α-plates grow inside the grain until they meet plates growing from other boundaries during further cooling phases. As a result, colonies of unidirectional α-plates are formed in the grain. Cutting the lamellas of different colonies and grains at the different angles in the plane of the microsample results in a visible difference in the α-plate thickness. The thicknesses of such plates are close to each other. The average microhardness of the buildup alloy was 397 HV0.3.

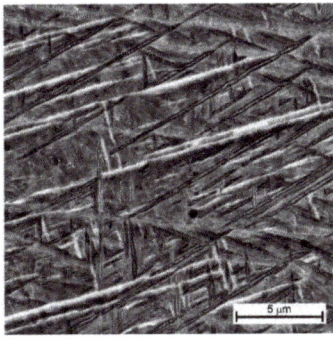

Figure 6. Microstructure of a Ti-6Al-4V buildup obtained by the DLD method.

3.2. Effect of the Temperature on the Fracture Behavior

A macrograph of the fracture surface of the specimens tested at different temperatures is shown in Figure 7. All of the specimens had a cup-and-cone ductile fracture. A distinctive feature of the specimens tested at 200 °C, as shown Figure 7a, is the presence of two zones: a fibrous zone and a shear zone. The fibrous zone corresponds to the area of slow crack growth. It is located in the center of the fracture. The shear zone is an annular fracture zone that is adjacent to the free surface of the specimen. The extent of the shear zone decreases until it disappears completely as test temperature increases. This is clearly visible in the macrographs of the specimens tested at 500 °C and 700 °C, shown in Figure 7b,c. The higher test temperature corresponds to a significantly higher reduction of the area and the presence of large pores with a diameter of approximately 350 μm.

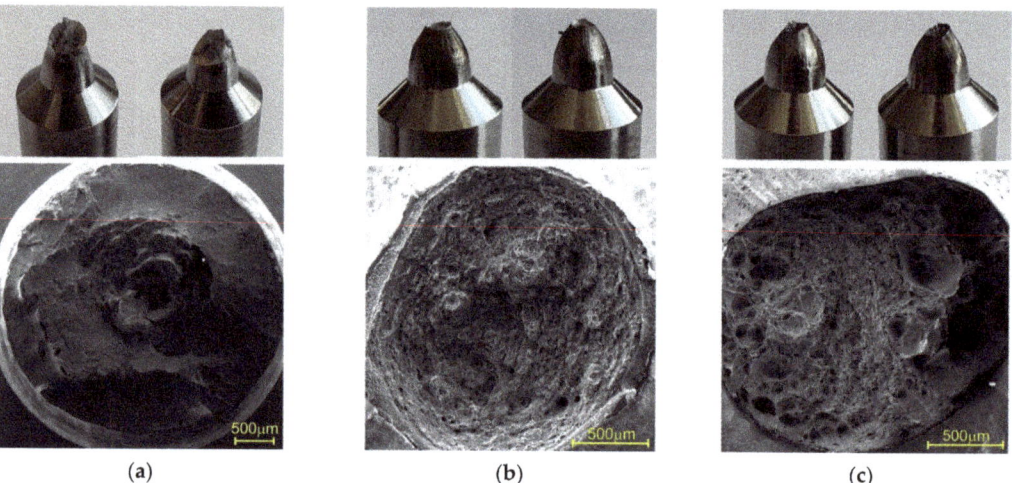

Figure 7. Fracture surface of specimens tested at (**a**) 200 °C, (**b**) 500 °C, and (**c**) 700 °C.

Deep dimples are clearly visible in the central region of the fracture surfaces of the specimens tested at 200 °C. A ductile local fracture is immediately initiated around those dimples, as seen in Figure 8a. The dimples are formed by the coalescence of micropores, which, in turn, grow and expand under a triaxial stress state [54,55]. Flat equiaxed dimples are clearly observed in the shear fracture zone shown in Figure 8b,c. They are formed due to the coalescence of micropores under the action of shear stresses. At higher test temperatures, the fracture is caused by the nucleation of the micropores at the grain boundaries, which are formed by a grain boundary slip, see Figure 9. The subsequent diffusion of vacancies or the development of local sliding leads to an enlargement of the pores. The larger dimples observed in Figure 9a correspond to triple grain boundary junctions. The small dimples seen in Figure 9b originate from the walls of the dislocation cells. At a test temperature of 700 °C, the pores are larger and deeper, as seen in Figure 9c. Note that the pores are elongated in the direction of plastic deformation.

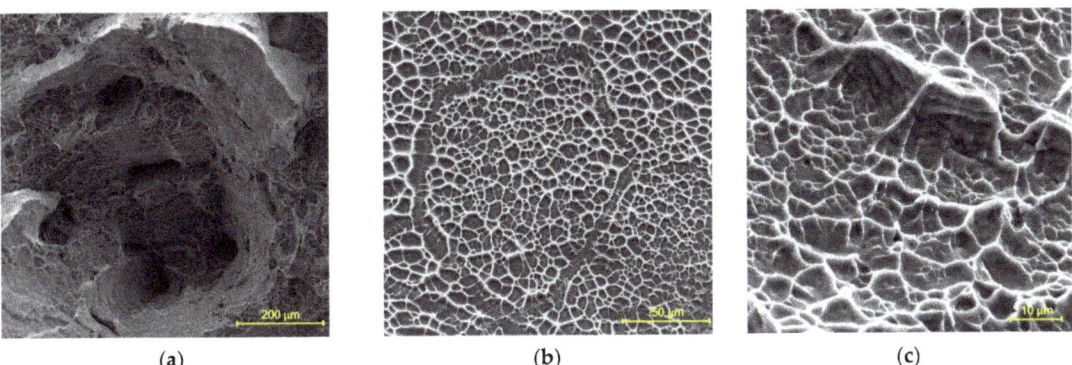

Figure 8. SEM fractograph of a specimen tested at 200 °C showing (**a**) the fibrous central zone and (**b**,**c**) the shear zone.

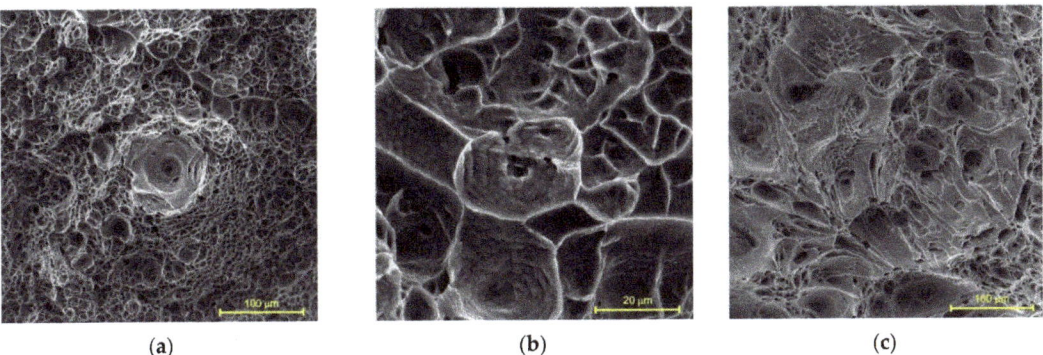

Figure 9. Fracture surface of specimen tested at (**a**,**b**) 500 °C and (**c**) 700 °C.

3.3. Short-Term Mechanical Properties of the Ti-6Al-4V Alloy over a Wide Temperature Range

The experimentally obtained engineering and true tensile stress curves of the Ti-6Al-4V alloy for the temperature range between 20 °C and 800 °C are shown in Figure 10. The processing of the experimental data was conducted according to the procedure described in Section 2.3. It can be seen in Figure 10a that the total strain corresponding to the ultimate strength increases when the temperature increases. The ductility of the alloy increases significantly at temperatures above 700 °C. It should be noted that the ductility of the DLD-processed alloy at room temperature is comparable to that of the wrought alloy [13].

In Figure 11a the yield and tensile strength are plotted as functions of the temperature. It can be observed that the yield strength decreases gradually by approximately 40% as the temperature rises to 500 °C. A further increase of the temperature leads to a significant increase in the softening rate. This behavior is associated with the intensification of the diffusion-controlled decomposition of the metastable α'-phase and the grain boundary slip process [56]. Thus, the yield strength decreases almost linearly from 600 MPa to 70 MPa in the temperature range between 500 °C and 800 °C. According to the published data shown in Figure 11b, a further increase in the temperature leads to complete softening of the material. The green curve corresponds to the sample with the α'-microstructure [29]. This shows a close correlation to the obtained curve for the DLD-processed alloy, especially at temperatures above 400 °C. The discrepancy between the curves in the temperature range of 20–400 °C can be explained by the differences in the size and the shape of the prior β-grain as well as by the morphology and the thickness of the α-plates [12]. On the other hand, the blue curve corresponds to the ($\alpha + \beta$) microstructure [31] and shows lower yield strength values. However, its behavior is almost identical to that of the obtained curve for the DLD-processed alloy.

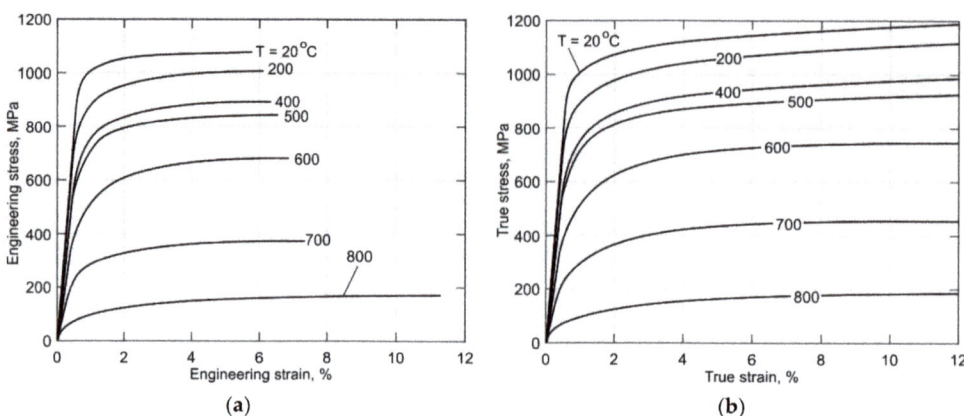

Figure 10. (a) Engineering and (b) true tensile curves of the DLD-processed Ti-6Al-4V alloy.

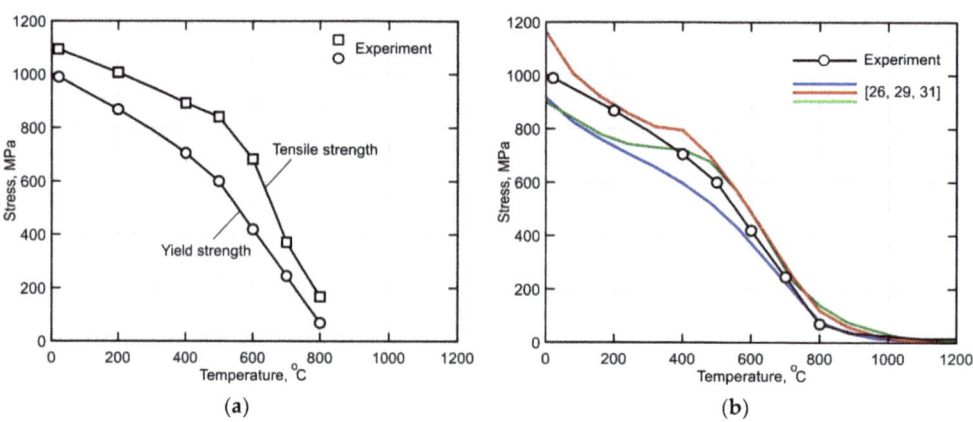

Figure 11. (a) Yield and tensile strength of the Ti-6Al-4V alloy as functions of temperature and (b) comparison of the obtained yield strength with published data from [26,29,31].

In Figure 12, the temperature-dependent Young's modulus is shown. It is observed that the Young's modulus remains almost unchanged for a temperature increase of up to 500 °C. However, it decreases sharply by approximately 70% from 109 GPa to 26 GPa upon further heating to 800 °C. A comparison of the obtained curves with previously published data shows a significant discrepancy. The blue curve corresponds to a sample with a bi-modal Widmanstätten microstructure obtained from a plate that was 12 mm thick. Note that the plates were treated by annealing for 6 h at 790 °C [31]. The DLD-processed alloy shows a Young's modulus with greater thermal stability. At 500 °C the Young's modulus of the alloy is about 46% higher than that of the wrought alloy.

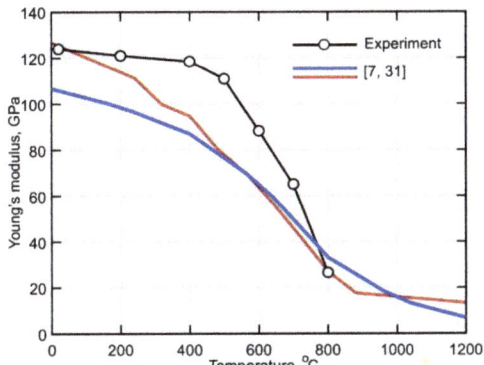

Figure 12. Comparison of the measured Young's modulus of the Ti-6Al-4V alloy with published data from [7,31].

An approximation of the obtained tensile curves for given temperatures using the proposed fitting function was performed according to the procedure described in Section 2.3. The obtained coefficients of the fitting function are given in Table 1. These data are of high importance for the numerical analysis of the residual stresses and distortion of additively manufactured parts.

Table 1. Mechanical properties and fitting coefficients.

T, °C	$\sigma_{0.2}$, MPa	E, GPa	True Tensile Curve			Engineering Tensile Curve		
			p_1	p_2	p_3	p_1	p_2	p_3
20	1000	124.0	0.49292	1.16000	0.012126	−0.34052	1.110000	0.0100000
200	845.0	121.0	0.49292	1.30004	0.012126	−0.34052	1.245515	0.0096846
400	704.8	118.5	0.47779	1.37959	0.009724	−0.33283	1.328941	0.0087338
500	600.0	110.9	0.39231	1.52868	0.007098	−0.64584	1.483647	0.0064257
600	418.8	88.2	−0.69472	1.93472	0.012777	−1.49563	1.842861	0.0115448
700	245	65.0	−1.5021	2.15010	0.018110	−2.46304	1.806981	0.0165517
800	70	26.5	−2.11940	3.33223	0.028902	−2.71973	3.036418	0.0238623

3.4. Temperature Dependence of the Thermal Expansion Coefficient

The experimentally obtained temperature dependence of the thermal strain is shown in Figure 13. It is not difficult to see that the heating and cooling parts of the curve have different slopes for temperatures above 600 °C. The temperature dependence of the thermal expansion coefficient was obtained according to the method described in detail in Section 2.4. A decrease of about 20% in the CTE occurs in the temperature range between 400 °C and 600 °C during heating, as seen in Figure 14a. According to [29], this can be explained by the diffusion-controlled phase transformation $\alpha' \rightarrow \alpha + \beta$, which is accompanied by a slight volume decrease. Note, that above 800 °C, the $\alpha + \beta \rightarrow \beta$ transformation begins and that the transformation rate is not constant. However, the transformation rate is rather slow in the interval between 800–900 °C, which is clearly visible in the secant CTE curve shown in Figure 14b. An increase of the diffusion mobility of the atoms at temperatures above 900 °C leads to a sufficient increase in the rate of β-phase formation. During holding at 1050 °C, the β-phase content reaches 100%, which leads to a reduction of the sample's volume. However, the coefficient of thermal expansion does not change significantly during cooling.

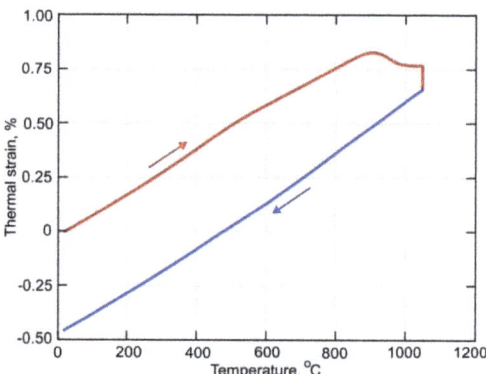

Figure 13. Temperature-dependent thermal strain curve of the DLD-processed Ti-6Al-4V alloy.

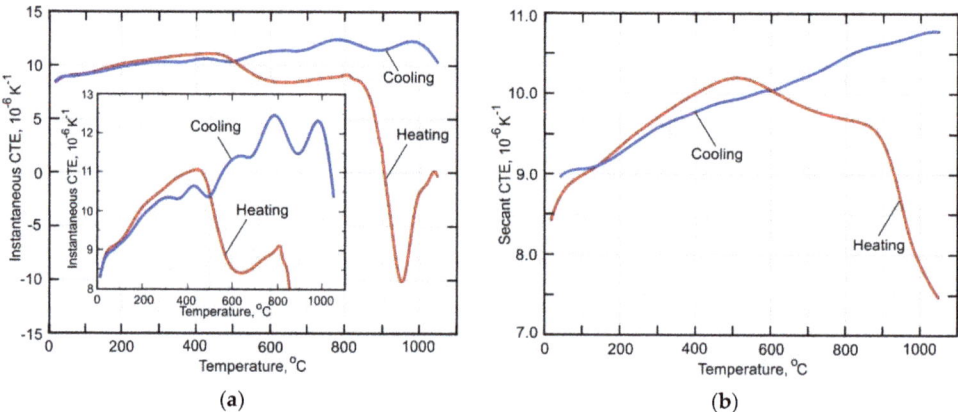

Figure 14. Temperature dependent coefficient of thermal expansion of (a) an instantaneous and (b) secant.

3.5. Analysis of the Stress Relaxation

The instability of the phase composition of the material and the relaxation of the residual stresses arising in the parts due to various technological operations may cause spontaneous changes of their size and shape over time, affecting their service properties. The conditions for relaxation are described by the following equation:

$$\varepsilon_0 = \varepsilon^e + \varepsilon^p = \frac{\sigma}{E} + \varepsilon^p = const \text{ at } \varepsilon^e \neq const; \ \varepsilon^p \neq const, \quad (4)$$

where ε_0 is the initial total strain, ε^e is the elastic strain, and ε^p is the plastic strain.

The total strain during the stress relaxation test remains constant due to the increase in plastic strain over time caused by the decrease of the fraction of the elastic strain. These processes can have a considerable effect on the shape stability of the part during DLD as well as during service. The experimentally obtained stress relaxation curves for 500 °C and 700 °C are shown in Figure 15. Note that for clarity, only a small part of the recorded experimental points is shown, as the data recording frequency was 500 Hz. The experimental data were approximated according to the following equation, which describes the intergranular diffusion relaxation processes [57]:

$$\sigma = \sigma_0 \cdot \exp\left(-\frac{k \cdot t}{1 + p \cdot t}\right), \qquad (5)$$

where σ_0 is the applied stress, and k and p are coefficients dependent on the temperature, the microstructure, and the phase composition.

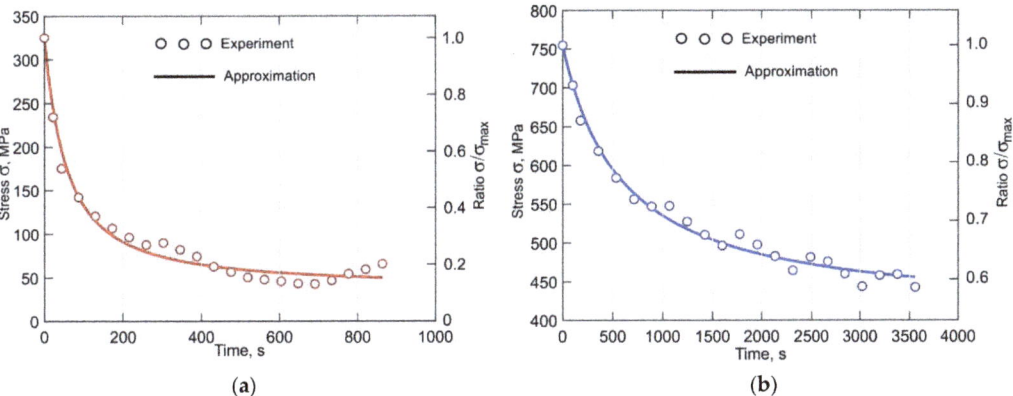

Figure 15. Stress relaxation curves of the DLD-processed Ti-6Al-4V alloy for (**a**) 700 °C and (**b**) 500 °C.

To determine the instantaneous creep strain rate, the time derivative of Equation (5) is obtained:

$$\dot{\varepsilon}_c = \frac{d\varepsilon_c}{dt} = -\frac{1}{E}\frac{d\sigma}{dt}, \qquad (6)$$

where E is the Young's modulus at a given temperature.

By differentiating Equation (5) and substituting it into Equation (6) the instantaneous creep strain rate is obtained:

$$\dot{\varepsilon}_c = \frac{\sigma_0}{E} \cdot \exp\left(-\frac{k \cdot t}{1 + p \cdot t}\right) \cdot \left[\frac{k}{(1 + p \cdot t)^2}\right]. \qquad (7)$$

It is a well-known fact that stress relaxation is a thermally activated process, which is particularly effective at high temperatures. Two regions can be distinguished in the curves shown in Figure 15. The first is characterized by an abrupt stress drop, and the second is characterized by a slow stress drop. As noted in [56], the sharp stress decrease at the beginning of the relaxation process is associated with the elimination of a large number of lattice defects. Over time, the amount of lattice distortions decreases, causing the relaxation rate to become slower. In addition, it can as well be explained by the fact that at the beginning of the relaxation phase, the value of the applied stress is high and thus closer to the yield strength of the individual crystallites and mosaic blocks. In the second region, the relaxation curve is asymptotic to a straight line that is parallel to the abscissa axis and shifts from it by the value of the peak stress at which relaxation will not occur. The kinetics of these processes are well illustrated by the creep rate curves shown in Figure 16. The creep rate at 700 °C is 0.12×10^{-3} % s^{-1} at the beginning of the relaxation process, which is 23 times higher than the creep rate at 500 °C, as seen in Figure 15a. A 50% reduction of the stress occurs during the first 60 s at 700 °C. After a sharp decrease, the stress continues to decrease, but at a considerably lower rate. In the first 600 s, the stress is reduced by a total of 86% at 700 °C and by 25% at 500 °C, as seen in Figure 15. Hence, it can be concluded that the overheating of the buildup due to its size and/or absence of inter-pass dwell time will lead to a significant reduction of the residual stresses. Therefore, published data on experimentally measured residual stresses without a detailed description of the process parameters affecting the temperature field cannot be used to analyze and verify the accuracy of simulation procedures.

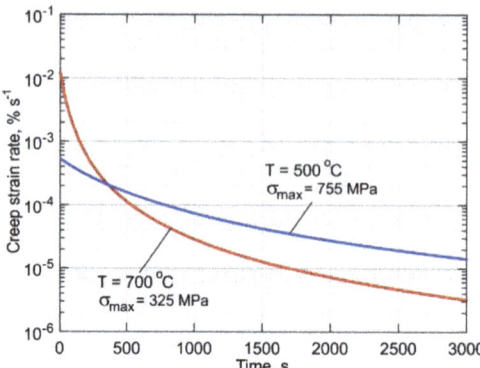

Figure 16. Creep rate curves of the DLD-processed Ti-6Al-4V for 500 °C and 700 °C.

4. Conclusions

The mechanical properties of the DLD-processed Ti-6Al-4V alloy were obtained by a tensile test performed in the temperature range between 20 °C and 800 °C. The influence of the thermal history on the stress relaxation for the cases with maximum temperatures of 500 °C and 700 °C were studied. In addition, the temperature dependence of the coefficient of thermal expansion was obtained. The influence of the initial microstructure of the samples on the deformation and the fractures at elevated temperatures was analyzed. An approximation of the measured tensile curves for given temperatures using a proposed fitting function was performed to describe the hardening behavior during plastic deformation. The following conclusions are drawn:

1. The microstructure of the buildup obtained by direct laser deposition with inter-pass temperatures in the range of 60–80 °C consists of a lamellar α'-phase and a small amount of residual β-phase.
2. According to the obtained stress curves, the yield strength decreases gradually by approximately 40% when the temperature increases to 500 °C. Furthermore, it was determined that the softening rate increases significantly upon further heating
3. It was found that the DLD-processed Ti-6Al-4V alloy has a Young's modulus with greater thermal stability than conventionally processed alloys. At 500 °C, the Young's modulus of the alloy is about 46% higher than that of the wrought alloy.
4. The analysis of the CTE curves showed that a diffusion-controlled transformation of $\alpha' \rightarrow \alpha + \beta$ in the temperature range between 400 °C and 600 °C leads to a 20% decrease in the CTE. In addition, the $\alpha + \beta \rightarrow \beta$ transformation was determined to start at temperatures above 800 °C.
5. The stress relaxation process was found to have a decisive influence on the formation of the residual stresses at temperatures above 700 °C, which is especially important in the production of small-sized parts by the DLD method.

Author Contributions: Conceptualization, S.I., M.G., and A.A.; methodology, S.I., A.A., M.G., and M.K.; formal analysis, investigation, S.I., M.G., and M.K.; data curation, S.I., A.A., and M.K.; writing—original draft preparation, S.I., M.G., and A.A.; writing—review and editing, S.I., M.G., A.A., and M.K.; visualization, S.I.; project administration, E.Z.; funding acquisition, E.Z. All authors have read and agreed to the published version of the manuscript.

Funding: This research was funded by the Ministry of Science and Higher Education of the Russian Federation as part of the World-class Research Center program: Advanced Digital Technologies (contract No. 075-15-2020-903 dated 16 November 2020).

Institutional Review Board Statement: Not applicable.

Informed Consent Statement: Not applicable.

Conflicts of Interest: The authors declare no conflict of interest. The funders had no role in the design of the study; in the collection, analyses, or interpretation of data; in the writing of the manuscript; or in the decision to publish the results.

References

1. DebRoy, T.; Wei, H.L.; Zuback, J.S.; Mukherjee, T.; Elmer, J.W.; Milewski, J.O.; Beese, A.M.; Wilson-Heid, A.; De, A.; Zhang, W. Additive manufacturing of metallic components-Process, structure and properties. *Prog. Mater. Sci.* **2018**, *92*, 112–224. [CrossRef]
2. Mukherjee, T.; Zuback, J.S.; Zhang, W.; DebRoy, T. Residual stresses and distortion in additively manufactured compositionally graded and dissimilar joints. *Comput. Mater. Sci.* **2018**, *143*, 325–337. [CrossRef]
3. Blakey-Milner, B.; Gradl, P.; Snedden, G.; Brooks, M.; Pitot, J.; Lopez, E.; Leary, M.; Berto, F.; du Plessis, A. Metal additive manufacturing in aerospace: A review. *Mater. Des.* **2021**, *209*, 110008. [CrossRef]
4. Gouge, M.; Michaleris, P. *Thermo-Mechanical Modeling of Additive Manufacturing*, 1st ed.; Butterworth-Heinemann: Oxford, UK, 2017.
5. Papadakis, L. Experimental and computational appraisal of the shape accuracy of a thin-walled virole aero-engine casing manufactured by means of laser metal deposition. *Prod. Eng. Res. Devel.* **2017**, *11*, 389–399. [CrossRef]
6. Babkin, K.; Zemlyakov, E.; Ivanov, S.; Vildanov, A.; Topalov, I.; Turichin, G. Distortion prediction and compensation in direct laser deposition of large axisymmetric Ti-6Al-4V part. *Procedia CIRP* **2020**, *94*, 357–361. [CrossRef]
7. Mukherjee, T.; Zhang, W.; DebRoy, T. An improved prediction of residual stresses and distortion in additive manufacturing. *Comput. Mater. Sci.* **2017**, *126*, 360–372. [CrossRef]
8. Seshacharyulu, T.; Medeiros, S.C.; Frazier, W.G.; Prasad, Y.V.R.K. Microstructural mechanisms during hot working of commercial grade Ti–6Al–4V with lamellar starting structure. *Mater. Sci. Eng. A* **2002**, *325*, 112–125. [CrossRef]
9. Rangaswamy, P.; Choo, H.; Prime, M.B.; Bourke, M.A.; Larsen, J.M. High temperature stress assessment in SCS-6/Ti-6Al-4V composite using neutron diffraction and finite element modeling. In *International Conference on Processing & Manufacturing of Advanced Materials*; Los Alamos National Laboratory: Las Vegas, NV, USA, 2000.
10. Denlinger, E.R.; Michaleris, P. Effect of stress relaxation on distortion in additive manufacturing process modeling. *Addit. Manuf.* **2016**, *12*, 51–59. [CrossRef]
11. Lu, X.; Lin, X.; Chiumenti, M.; Cervera, M.; Li, J.; Ma, L.; Wei, L.; Hu, Y.; Huang, W. Finite element analysis and experimental validation of the thermomechanical behavior in laser solid forming of Ti-6Al-4V. *Addit. Manuf.* **2018**, *21*, 30–40. [CrossRef]
12. Tiley, J.S. Modeling of Microstructure Property Relationships in Ti-6Al-4V. Ph.D. Thesis, Ohio State University, Columbus, OH, USA, 2002.
13. Donachie, M.J. *Titanium: A Technical Guide*, 2nd ed.; ASM International: Materials Park, OH, USA, 1988.
14. Moiseyev, V.N. *Titanium Alloys: Russian Aircraft and Aerospace Applications*; CRC Press Taylor & Francis Group: Boca Raton, FL, USA, 2006.
15. Boyer, R.; Collings, E.W.; Welsch, G. *Properties Handbook: Titanium Alloys*; ASM International: Materials Park, OH, USA, 1994.
16. Mills, K.C. *Recommended Values of Thermophysical Properties for Selected Commercial Alloys*; Woodhead Publishing: Cambridge, UK, 2002.
17. Ilyin, A.A.; Kolachev, B.A.; Polkin, I.S. *Titanium Alloys. Composition, Structure, Properties. Handbook*; VILS-MATI: Moscow, Russia, 2009. (In Russian)
18. *ASTM F2924-14 Standard Specification for Additive Manufacturing Titanium-6 Aluminum-4 Vanadium with Powder Bed Fusion*; ASTM International: West Conshohocken, PA, USA, 2021.
19. Ding, R.; Guo, Z.X.; Wilson, A. Microstructural evolution of a Ti-6Al-4V alloy during thermomechanical processing. *Mater. Sci. Eng. A* **2002**, *327*, 233–245. [CrossRef]
20. Weiss, I.; Froes, F.H.; Eylon, D.; Welsch, D.G.E. Modification of alpha morphology in Ti-6Al-4V by thermomechanical processing. *Metall. Mater. Trans. A* **1986**, *17*, 1935–1947. [CrossRef]
21. Liu, S.; Shin, Y.C. Additive manufacturing of Ti6Al4V alloy: A review. *Mater. Des.* **2019**, *164*, 107552. [CrossRef]
22. Lewandowski, J.J.; Seifi, M. Metal Additive Manufacturing: A Review of Mechanical Properties. *Annu. Rev. Mater. Res.* **2016**, *46*, 151–186. [CrossRef]
23. Lin, Y.C.; Jiang, X.-Y.; Shuai, C.; Zhao, C.-Y.; He, D.-G.; Chen, M.-S.; Chen, C. Effects of initial microstructures on hot tensile deformation behaviors and fracture characteristics of Ti-6Al-4V alloy. *Mater. Sci. Eng. A* **2018**, *711*, 293–302. [CrossRef]
24. Paghandeh, M.; Zarei-Hanzaki, A.; Abedi, H.R.; Vahidshad, Y. On the warm temperature strain accommodation mechanisms of Ti-6Al-4V alloy holding different starting microstructures. *J. Mater. Res. Technol.* **2021**, *14*, 496–506. [CrossRef]
25. Rangaswamy, P. Comparison of residual strains measured by X-ray and neutron diffraction in a titanium (Ti–6Al–4V) matrix composite. *Mater. Sci. Eng. A* **1999**, *259*, 209–219. [CrossRef]
26. Zhao, X.; Iyer, A.; Promoppatum, P.; Yao, S.-C. Numerical modeling of the thermal behavior and residual stress in the direct metal laser sintering process of titanium alloy products. *Addit. Manuf.* **2017**, *14*, 126–136. [CrossRef]

27. Chiumenti, M.; Cervera, M.; Dialami, N.; Wu, B.; Jinwei, L.; Agelet de Saracibar, C. Numerical modeling of the electron beam welding and its experimental validation. *Finite Elem. Anal. Des.* **2016**, *121*, 118–133. [CrossRef]
28. Cao, J.; Gharghouri, M.A.; Nash, P. Finite-element analysis and experimental validation of thermal residual stress and distortion in electron beam additive manufactured Ti-6Al-4V build plates. *J. Mater. Process. Technol.* **2016**, *237*, 409–419. [CrossRef]
29. Robert, Y. Simulation Numérique du Soudage du TA6V par Laser YAG Impulsionnel: Caractérisation Expérimentale et Modélisation des Aspects Thermomécaniques Associés à ce Proceed. Ph.D. Thesis, École Nationale Supérieure des Mines de Paris, Paris, France, 2007. (In French)
30. MSC Software, Simufact Additive, Material Database. 2016. Available online: https://www.mscsoftware.com/product/simufact-additive (accessed on 28 September 2021).
31. Babu, B. Physically Based Model for Plasticity and Creep of Ti-6Al-4V. Ph.D. Thesis, Luleå University of Technology, Luleå, Sweden, 2008.
32. Babu, B.; Lundbäck, A.; Lindgren, L.-E. Simulation of Ti-6Al-4V Additive Manufacturing Using Coupled Physically Based Flow Stress and Metallurgical Model. *Materials* **2019**, *12*, 3844. [CrossRef]
33. Yang, Y.; Liu, Y.J.; Chen, J.; Wang, H.L.; Zhang, Z.Q.; Lu, Y.J.; Wu, S.Q.; Lin, J.X. Crystallographic features of α variants and β phase for Ti-6Al-4Valloy fabricated by selective laser melting. *Mater. Sci. Eng. A* **2017**, *707*, 548–558. [CrossRef]
34. Saboori, A.; Abdi, A.; Fetami, S.A.; Marchese, G.; Biamino, S.; Mirzadeh, H. Hot deformation behavior and flow stress modeling of Ti–6Al–4V alloy produced via electron beam melting additive manufacturing technology in single β-phase field. *Mater. Sci. Eng. A* **2020**, *792*, 139822. [CrossRef]
35. Bambach, M.; Sizova, I.; Szyndler, J.; Bennett, J.; Hyatt, G.; Cao, J.; Papke, T.; Merklein, M. On the hot deformation behavior of Ti-6Al-4V made by additive manufacturing. *J. Mater. Process. Technol.* **2021**, *288*, 116840. [CrossRef]
36. Song, J.; Han, Y.; Fang, M.; Hu, F.; Ke, L.; Li, Y.; Lei, L.; Lu, W. Temperature sensitivity of mechanical properties and microstructure during moderate temperature deformation of selective laser melted Ti-6Al-4V alloy. *Mater. Charact.* **2020**, *165*, 110342. [CrossRef]
37. Li, P.-H.; Guo, W.-G.; Huang, W.-D.; Su, Y.; Lin, X.; Yuan, K.-B. Thermomechanical response of 3D laser-deposited Ti-6Al-4V alloy over a wide range of strain rates and temperatures. *Mater. Sci. Eng. A* **2015**, *647*, 34–42. [CrossRef]
38. Motoyama, Y.; Tokunaga, H.; Kajino, S.; Okane, T. Stress-strain behavior of a selective laser melted Ti-6Al-4V at strain rates of 0.001–1/s and temperatures 20–1000 °C. *J. Mater. Process. Technol.* **2021**, *294*, 117141. [CrossRef]
39. *ASTM F136-02a Standard Specification for Wrought Titanium-6 Aluminum-4 Vanadium ELI (Extra Low Interstitial) Alloy for Surgical Implant Applications (UNS R56401)*; ASTM International: West Conshohocken, PA, USA, 2002.
40. *ASM Handbook, Metallography and Microstructures*; ASM Handbook series; ASM International: Materials Park, OH, USA, 2004; Volume 9.
41. Garcia, D. Robust smoothing of gridded data in one and higher dimensions with missing values. *Comput. Stat. Data Anal.* **2010**, *54*, 1167–1178. [CrossRef] [PubMed]
42. Oppenheim, A.V.; Schafer, R.W. *Discrete-Time Signal Processing*, 3rd ed.; Prentice-Hall: Hoboken, NJ, USA, 2010.
43. Dennis, J.E., Jr. Nonlinear Least-Squares. In *The State of the Art in Numerical Analysis*; Jacobs, D., Ed.; Academic Press: New York, NY, USA, 1977; pp. 269–312.
44. *ASTM E328-21, Standard Test Methods for Stress Relaxation for Materials and Structures*; ASTM International: West Conshohocken, PA, USA, 2021.
45. Wu, B.; Pan, Z.; Ding, D.; Cuiuri, D.; Li, H. Effects of heat accumulation on microstructure and mechanical properties of Ti6Al4V alloy deposited by wire arc additive manufacturing. *Addit. Manuf.* **2018**, *23*, 151–160. [CrossRef]
46. Foster, B.K.; Beese, A.M.; Keist, J.S.; McHale, E.T.; Palmer, T.A. Impact of Interlayer Dwell Time on Microstructure and Mechanical Properties of Nickel and Titanium Alloys. *Metall. Mater. Trans. A* **2017**, *48*, 4411–4422. [CrossRef]
47. Song, T.; Dong, T.; Lu, S.L.; Kondoh, K.; Das, R.; Brandt, M.; Qian, M. Simulation-informed laser metal powder deposition of Ti-6Al-4V with ultrafine α-β lamellar structures for desired tensile properties. *Addit. Manuf.* **2021**, *46*, 102139.
48. Xiao, Y.; Cagle, M.; Mujahid, S.; Liu, P.; Wang, Z.; Yang, W.; Chen, L. A gleeble-assisted study of phase evolution of Ti-6Al-4V induced by thermal cycles during additive manufacturing. *J. Alloys Compd.* **2021**, *860*, 158409. [CrossRef]
49. Li, X.; Tan, W. Numerical investigation of effects of nucleation mechanisms on grain structure in metal additive manufacturing. *Comput. Mater. Sci.* **2018**, *153*, 159–169. [CrossRef]
50. Lin, J.J.; Lv, Y.H.; Liu, Y.X.; Xu, B.S.; Sun, Z.; Li, Z.G.; Wu, Y.X. Microstructural evolution and mechanical properties of Ti-6Al-4V wall deposited by pulsed plasma arc additive manufacturing. *Mater. Des.* **2016**, *102*, 30–40. [CrossRef]
51. Klimova-Korsmik, O.G.; Turichin, G.A.; Shalnova, S.A.; Gushchina, M.O.; Cheverikin, V.V. Structure and properties of Ti-6Al-4V titanium alloy products obtained by direct laser deposition and subsequent heat treatment. *J. Phys. Conf. Ser.* **2018**, *1109*, 012061. [CrossRef]
52. Shalnova, S.A.; Klimova-Korsmik, O.G.; Turichin, G.A.; Gushchina, M. Effect of process parameters on quality of Ti-6Al-4V multi-layer single pass wall during direct laser deposition with beam oscillation. *Solid State Phenom.* **2020**, *299*, 716–722. [CrossRef]
53. Gushchina, M.O.; Ivanov, S.Y.; Vildanov, A.M. Effect of Temperature Field on Mechanical Properties of Direct Laser Deposited Ti-6Al-4V Alloy. *IOP Conf. Ser. Mater. Sci. Eng.* **2020**, *969*, 012103. [CrossRef]
54. Safyari, M.; Moshtaghi, M.; Kuramoto, S. Environmental hydrogen embrittlement associated with decohesion and void formation at soluble coarse particles in a cold-rolled Al-Cu based alloy. *Mater. Sci. Eng. A* **2021**, *799*, 139850. [CrossRef]
55. Davis, J.R. *Tensile Testing*, 2nd ed.; ASM International: Materials Park, OH, USA, 2004.

56. Kassner, M.E. *Fundamentals of Creep in Metals and Alloys*, 3rd ed.; Butterworth-Heinemann: Oxford, UK, 2015.
57. Oding, I.A. *Creep and Stress Relaxation in Metals*; Oliver & Boyd: Edinburgh, UK; London, UK, 1965.

Preparation of W-C-Co Composite Micropowder with Spherical Shaped Particles Using Plasma Technologies

Andrey Samokhin *, Nikolay Alekseev, Aleksey Astashov, Aleksey Dorofeev, Andrey Fadeev, Mikhail Sinayskiy and Yulian Kalashnikov

Baikov Institute of Metallurgy and Materials Science of the Russian Academy of Sciences, 49, Leninskiy prosp., 119334 Moscow, Russia; nvalexeev@yandex.ru (N.A.); aastashov@imet.ac.ru (A.A.); adorofeev@imet.ac.ru (A.D.); afadeev@imet.ac.ru (A.F.); sinaisky@imet.ac.ru (M.S.); ulian1996@inbox.ru (Y.K.)
* Correspondence: asamokhin@imet.ac.ru

Abstract: The possibility of obtaining composite micropowders of the W-C-Co system with a spherical particle shape having a submicron/nanoscale internal structure was experimentally confirmed. In the course of work carried out, W-C-Co system nanopowders with the average particle size of approximately 50 nm were produced by plasma-chemical synthesis. This method resulted in the uniform distribution of W, Co and C among the nanoparticles of the powder in the nanometer scale range. Dense microgranules with an average size of 40 microns were obtained from the nanopowders by spray drying. The spherical micropowders with an average particle size of 20 microns were received as a result of plasma treatment of 25.36 microns microgranule fraction. The spherical particles obtained in the experiments had a predominantly dense microstructure and had no internal cavities. The influence of plasma treatment process parameters on dispersity, phase, and chemical composition of spherical micropowders and powder particles microstructure has been established.

Keywords: tungsten carbides; cobalt; nanopowder; synthesis; granulation; spheroidization; DC thermal plasma

1. Introduction

Hard metals based on tungsten carbide are widely used in the production of cutting tools for metalworking, tools for rock drilling, wear-resistant parts, and coatings, etc. [1–3].

Hard metal products are manufactured by various methods, including powder metallurgy methods, but the production of complex-shaped parts encounters significant difficulties. Nowadays, intensively developing additive technologies make it possible to overcome the problems of manufacturing parts with complex shapes, so the attention of researchers and developers is drawn to the development of additive technologies for the manufacture of hard metal compacts. An overview of current research in this area can be found in [4].

Metal powders used in the layer-by-layer growth of products by methods of additive technologies should have good flowability and provide the highest possible packing density of particles when creating a powder layer [5,6]. These requirements may be achieved by using powders with a spherical particle shape having a size in the range from 20 to 60 μm.

An effective method for producing powders with a spherical particle shape is the processing of powders consisting of particles with an irregular shape in a flow of thermal plasma of electric discharges where the particles melt and become spherical due to surface tension forces. Plasma processes of spheroidization, the studies of which were started in the second half of the last century [7,8], are now widely used for processing metal powders [9–14]. However, the production of powders of WC-Co hard metals with a spherical particle shape in plasma processes has been studied only in a very limited number of works.

The authors of [15] obtained spherical powders by processing composite microgranules WC-12 wt.% Co in a thermal plasma flow generated by an electric arc plasma torch. The initial microgranules had an average size of 38 µm and were obtained by spray drying a suspension of WC and Co powders. Plasma treatment of microgranules made it possible to reduce the porosity in composite microparticles and ensure their spheroidization; however, high-temperature treatment led to undesirable effects—the transformation of some parts of WC carbide into W_2C and the partial evaporation of cobalt. The size of WC grains in the obtained spherical particles was in the micron range.

The authors of the paper [16] showed the possibility of obtaining dense spherical microparticles of the hard alloy WC-Co as a result of processing the corresponding porous microgranules in a thermal plasma flow followed by a heat treatment. Initial WC-30 wt.% Co microgranules with an average size of 87 µm were obtained by spray drying a suspension of a mixture of initial WC and Co powders. As a result of the processing of porous microgranules, dense spherical particles with a changed phase composition, represented by the phases C, Co, W_2C, and Co_3W_3C, were obtained. Subsequent heat treatment of this powder at a temperature of 950 °C provided a return to the original phase composition WC-Co, while the particles retained their spherical shape, and the powder had the fluidity required for its use in additive technologies.

In the above-mentioned works [15,16], the obtained spherical WC-Co particles had a microstructure with micron-sized WC grains.

The particle size of tungsten carbide powder is one of the critical factors determining the mechanical properties of WC-Co hard metals and the transition to nanostructured hard metals is a means to significantly improve these properties [17,18].

The aim of this research was to experimentally determine the possibility of obtaining composite W-C-Co system micropowders with a spherical particle shape having a submicron/nanoscale internal structure, using an approach involving three successive stages:

- W-C-Co system nanopowders synthesis by interaction of tungsten oxide WO_3 and cobalt powder mixture with methane in a flow of hydrogen-containing thermal plasma of electric arc plasm torch.
- Nanopowders granulation in a spray dryer to produce nanopowder microgranules, and classification of microgranules into a given fraction.
- Densification and spheroidization in a thermal plasma of the separated fractions of microgranules.

In some cases, the final classification of dense spherical particles obtained after plasma treatment is required with the purpose of removing nano- and submicron particles formed during condensation of vaporized material.

The proposed approach can make it possible to obtain micropowders of a WC-Co hard metal with a spherical particle shape and a submicron microstructure. To date, such powders have not been produced, but they are of interest for the manufacture of hard metal parts of complex shapes with an ultrafine microstructure using modern methods of additive technologies.

2. Materials and Methods

2.1. Obtaining Nanopowder of W-C-Co System

W-C-Co system nanopowder was obtained by processing of mixture of tungsten oxide WO_3 and cobalt Co powders with methane in thermal plasma flow generated in electric arc plasma torch with self-adjusting arc length with nominal power of 30 kW. Detailed description of the setup and processes of nanoparticles formation in the thermal plasma flow are presented in [9].

Powder mixture consisted of tungsten oxide WO_3 particles with average particle size less than 50 µm and Co particles with particle size less than 5 µm. All experiments were performed at a constant mass ratio of elements W/Co = 11 in the raw mix.

Nitrogen, as a part of plasma-forming and transport gases, was supplied from the air separation unit, the oxygen content in nitrogen was no more than 0.01 vol.%. Hydrogen, with purity not less than 99.95 vol.%, was added to plasma-forming gas to ensure reduction of tungsten oxide. Compressed methane, with a purity of 99.9 vol.%, was used as a carbidizer.

2.2. W-C-Co System Nanopowder Microgranules Production

Nanopowder microgranules of W-C-Co system were obtained by spray drying of aqueous suspensions of W-C-Co nanopowder on Buchi Mini Spray Dryer B-290 (Flawil, Switzerland) equipped with an ultrasonic nozzle.

The production of microgranules of the W-C-Co system based on spray drying includes three main stages:

- Preparation of an alcohol suspension consisting of composite nanoparticles of the W-C-Co system and polyvinyl butyral $(C_8H_{14}O_2)_n$ (PVB) used as an organic binder to ensure the strength of the obtained microgranules.
- Spray drying of the obtained suspension with an ultrasonic nozzle. Nitrogen was used as a working gas in the process of granulating the nanopowder using a Buchi B-295 circulation gas circuit.
- Separation of the 25.63 μm fraction of microgranules on a Retsch AS 200 sieve machine (Haan, Germany).

2.3. Plasma Processing of W-C-Co Nanopowders

Separated fraction of nanopowder granules was treated in a thermal argon-hydrogen (3 vol.%) plasma jet, generated in an electric arc plasma torch with a rated power of 30 kW. A detailed description of the plasma setup for powders spheroidization used is given in [10]. The formation of nano- and submicron particles formed due to partial evaporation and subsequent condensation of a raw material. Destruction of some granules by collision and removal of thermal gasification products of an organic binder from granules enhances the evaporation processes during plasma treatment of nanopowder microgranules. Sedimentation in a distilled water after ultrasonic dispersion was used to remove nano- and submicron particles from plasma treated powder. In this process, spheroidized microparticles are separated from the resulting suspension by sedimentation and subsequent drying in vacuum, while nano- and submicron particles are removed with water as a suspension.

A comprehensive analysis of the physicochemical properties of powders included:

- Scanning, transmission electron and optical microscopy—Scios SEM microscope (FEI, Hillsboro, OR, USA) with elemental energy dispersive X-ray microanalysis (EDS), Osiris TEM microscope (FEI, Hillsboro, OR, USA) and Olympus CX31 OM microscope (Tokyo, Japan), respectively. ImageScope M software (Aperio Technologies, Vista, CA, USA) was used for statistical image processing.
- Measurement of the specific surface area of nanopowders by the BET method on a Micromeritics TriStar 3000 specific surface analyzer (Norcross, GA USA).
- Amount of total oxygen measurements with a TC-600 (LECO, St. Joseph, MO, USA) analyzer during reduction smelting in a graphite crucible and detection of the resulting gases with an infrared radiation sensor.
- Amount of total carbon in a CS-600 (LECO, St. Joseph, MO, USA) analyzer by burning the sample in a stream of oxygen and detecting the resulting gases using an infrared radiation sensor.
- Determination of metallic elements (Co, W) by X-ray fluorescence spectroscopy (XRFMS) in the powder layer on an Orbis analyzer (EDAX, Mahwah, NJ, USA).
- Particle size distribution of micropowders measured with a Mastersizer 2000M laser diffraction particle size analyzer (Malvern, Worcestershire, UK) with Hydro S liquid sample feeder.
- Phase analysis on an Ultima-4 X-ray diffractometer (RIGAKU, Tokyo, Japan) with a monochromator in filtered Cu-Kα radiation.

- Separation of nanoparticles in spheroidal micropowders by fractional separation in liquid by sedimentation of aqueous suspension after treatment on ultrasonic dispersant Bandelin Sonopuls HD3100 (Berlin, Germany).
- Determination of flowability of powders was carried out using a calibrated funnel (Hall device) and a stopwatch for samples weighing 50 g.
- Determination of the apparent density of powders by weight method using a funnel in accordance with GOST 19440-94 [19] (ISO 3923-1:2018-09 [20], ISO 3923-2:1981 [21]).

3. Results and Discussion

3.1. Preparation of W-C-Co System Nanopowders

Plasma-chemical synthesis of W-C-Co system nanopowders was carried out in the plasma reactor with the confined jet flow. The power of the plasma torch was in the range of 19.24 kW. A mixture of nitrogen and hydrogen (20 vol.%) was used as plasma-forming gas. Dispersed raw material was transported with a feeding rate of 5 g/min by a hydrogen–methane mixture.

According to the results of electron microscopy (Figure 1), obtained nanopowders represent a polydisperse system consisting of aggregated nanoparticles in the size range from 10 to 100 nm, with the nanoparticles predominantly close to spherical in shape.

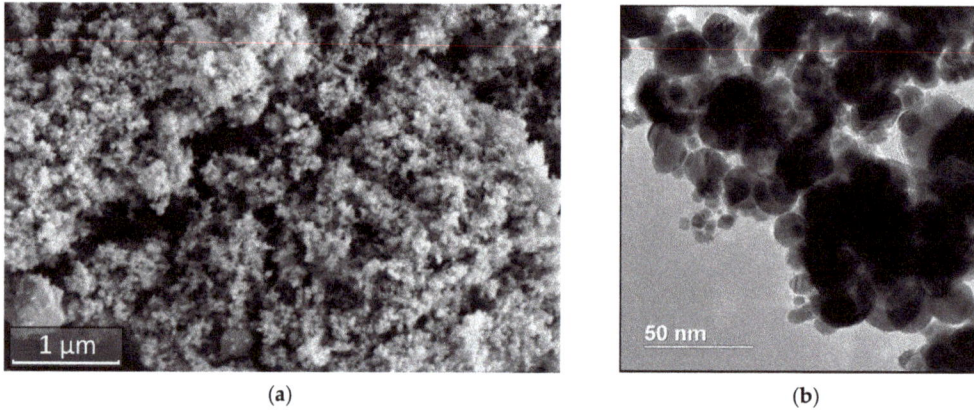

Figure 1. SEM (**a**) иTEM (**b**) images of W-C-Co nanopowder.

The specific surface area of nanopowders obtained ranged from 15 to 20 m^2/g. The main process parameters affecting the values of the specific surface area are the feed rate of dispersed raw materials and the methane flow rate. With increasing feed rate of dispersed raw materials, the specific surface area decreases due to a boost in the concentration of condensed vapors, and an increase in methane flow rate leads to a boost in the specific surface area of the nanopowder due to an increase in the content of free carbon with a high specific surface area.

Phase composition of the obtained W-C-Co system nanopowder containing 4.7 wt.% of carbon is characterized by the predominance of the W$_2$C tungsten carbide phase with the presence of WC and W phases (Figure 2a).

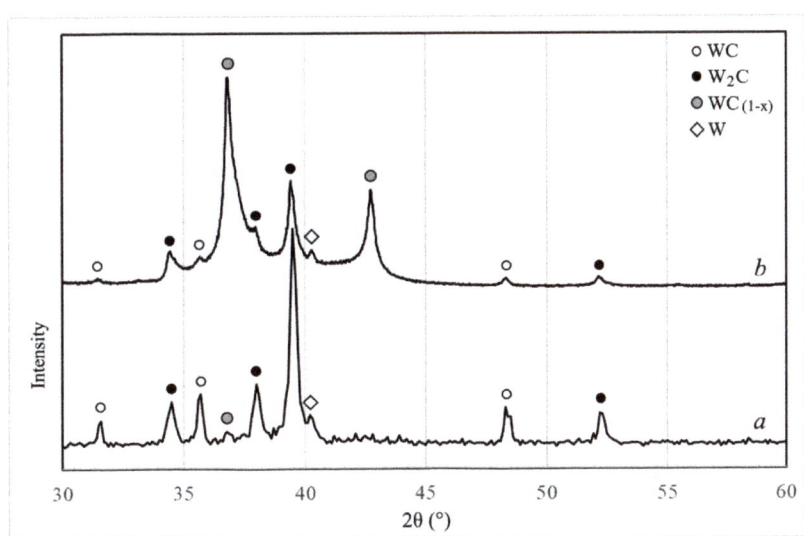

Figure 2. XRD patterns of W-C-Co (**a**) and W-C (**b**) nanopowders produced by plasma synthesis.

It is noted that during the plasma-chemical synthesis of nanopowders of the W-C and W-C-Co systems, the phase composition of the resulting nanopowders is noticeably different (Figure 2). Introduction of a cobalt into the process leads to a significant decrease in a cubic tungsten carbide phase WC_{1-x} and an increase in hexagonal tungsten monocarbide phase WC content as well as an increase in W_2C phase.

The results of the EDS microanalysis indicate that the elements W, Co, and C are evenly distributed among the nanoparticles of the powder of the W-C-Co composition with a uniformity scale in the nanometer size range (Figure 3). The high level of uniformity in the elemental composition of the obtained nanopowder is determined by the mechanism of its formation based on the co-condensation of components from the gas phase. Therefore, when carrying out this process, special attention was paid to creating structural and technological conditions to ensure the most complete evaporation of the initial disperse raw materials and the implementation of the targeted chemical reactions. The yield of the W-C-Co composition nanopowder reached 98.0–99.5 wt.%. The content of micron-sized particles in the nanopowders did not exceed 0.5–1.5 wt.%. We attribute their content mainly to incomplete evaporation of the raw material particles. The oxygen content in this fraction was 3–5 wt.%.

For the following stages of nanopowder granulation and microgranule spheroidization, a batch of W-C-Co system nanopowder containing 7.7 wt.% cobalt and 4.7 wt.% carbon was produced. Total oxygen content of the powder was 0.5 wt.%. The carbon content in the nanopowder was reduced in relation to the value corresponding to tungsten monocarbide WC since an organic binder was used for making microgranules which pyrolysis during plasma treatment leading to the formation of carbon in the microgranule volume.

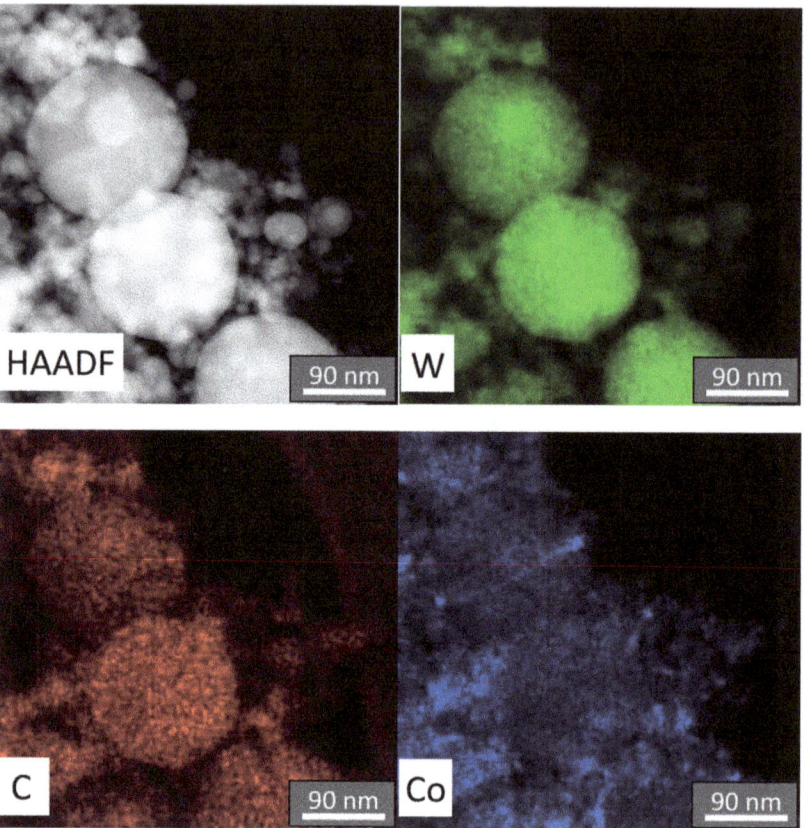

Figure 3. Results of the EDS microanalysis with element distribution maps (W, C, Co) among particles of W-C-Co composite nanopowder.

3.2. Preparation of W-C-Co microgranules

The granulation process of the obtained nanopowders was carried out in a spray drying unit at drying gas temperatures of 40–150 °C with a flow rate of 20 m^3/h. The nozzle cooling gas flow rate was 0.3 m^3/h. Flow rate of the suspension was in the range from 3 to 12 g/min. The power of the ultrasonic nozzle in all cases was 3 W.

The microgranules obtained in the spray drying process were subjected to sieving with the extraction of fractions of 25–63 microns, the yield of which was about 50%. Microgranules have mostly irregular shapes (Figure 4), determined by drying conditions, as well as the use of a low-boiling liquid-ethanol—as a dispersion medium. When distilled water and water-soluble organic binder were used as a dispersion medium we obtained spherical granules, but their strength was insufficient, and the yield of the target fraction was low. This is largely due to the fact that nanopowders of the W-C-Co system have a pyrocarbon layer on the surface that is formed as a result of hydrocarbons pyrolysis [22] and, therefore, the water suspensions of the W-C-Co nanopowder system have a poor stability.

The dispersed composition of separated granule fraction was characterized by the values of distribution parameters D_{10} = 23 μm, D_{50} = 37 μm, and D_{90} = 60 μm (Figure 5). The average particle size of microgranules was 39 μm.

Figure 4. SEM image of W-C-Co microgranules obtained in the process of spray drying.

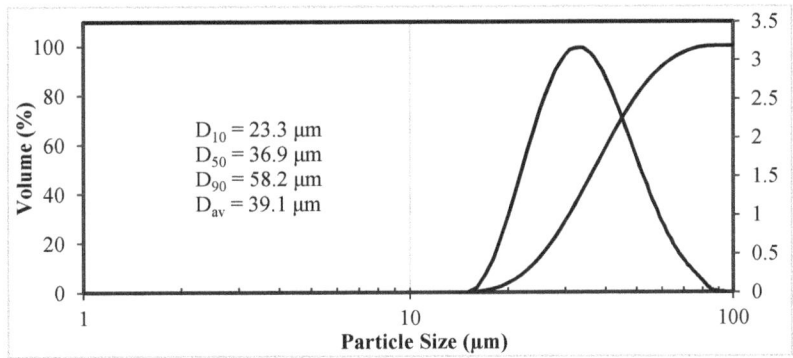

Figure 5. Differential and integral microgranules size.

According to the analysis results, the carbon content in the microgranules was 6.4 wt.%, oxygen-1.0 wt.%, the apparent density of the microgranules was 2.6 g/cm^3, and the flowability was 29 s/50 g. An increase in carbon content by 1.7 wt.% and oxygen content by 0.5 wt.% was due to the use of an organic binder. An additional channel for increasing oxygen in the granules was the interaction of nanoparticles with active oxygen-containing components formed during ultrasonic dispersion of an alcohol-based suspension at the stage of preparing a suspension for spray drying.

Selected granulation mode in combination with the use of an alcohol-based dispersion medium and polyvinyl butyral made it possible to ensure the strength of microgranules sufficient for their transportation without destruction from the powder feeder into the thermal plasma flow by the carrier gas.

3.3. Plasma Treatment of W-C-Co System Microgranules

Fraction of nanopowders with a size range of 25 to 63 microns containing 6.4 wt.% carbon was treated in a thermal plasma jet with a mixture of Ar + 5 vol.% H_2. Use of hydrogen in the plasma-forming gas allowed to increase the intensity of heating of granules due to its high thermal conductivity and, as a result, the degree of spheroidization of the obtained particles.

At a plasma-forming gas flow rate of 2.0 m^3/h, the power input of the plasma torch in the conducted experiments was 20–30 kW. The enthalpy of the plasma jet varied in the range from 2.4 to 4.9 kW·h/m^3. Microgranules with a flow rate of 6 g/min were transported from the feeder to the plasma reactor with argon at a flow rate of 0.5 m^3/h.

In all experiments performed, within the specified range of process parameters variation, the degree of spheroidization of microgranules was at least 90% (Figure 6a). Apparent density of obtained spheroidized powder (after removal of nano- and submicron particles) changed from 8.8 to 9.6 g/cm^3 with an increasing enthalpy of the plasma jet (Table 1).

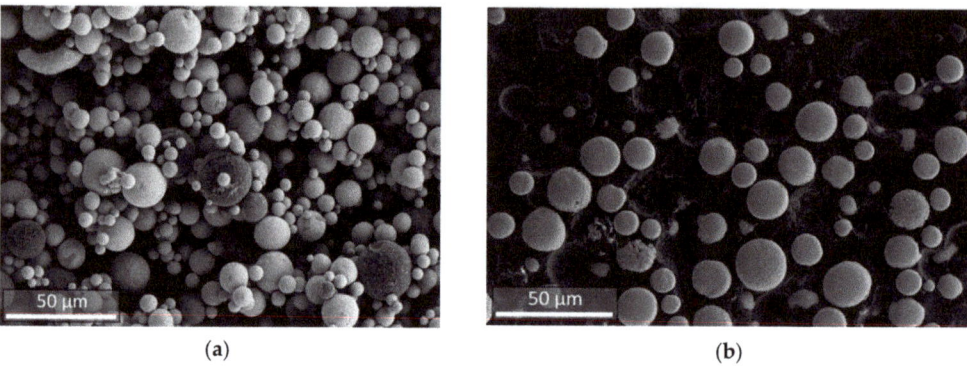

(a)   (b)

Figure 6. SEM images of W-C-Co micropowder after the spheroidization and removal of nano- and submicron particles (a) and metallographic cross-section of this powder (b).

Table 1. Characteristics of W-C-Co powders.

	Microgranules	Spheroidized Powder Produced at Different Enthalpy	
		2.8 kW·h/m^3	4.8 kW·h/m^3
D_{av}, μm	39.1	16.7	19.8
Apparent density, g/cm^3	2.6	8.8	9.5
Flowability, s/50 g	29	11	10

The experiments performed showed that an increase in the enthalpy of the plasma jet in the range from 2.4 to 4.9 kW h/m^3 leads to nanoparticles content increasing in the treated powder from 8 to 13 wt.%. The formation of nanoparticles occurs due to the evaporation and subsequent condensation of microgranule components. The elemental composition of the nanoparticles is characterized by a cobalt content of 70 wt.% and a tungsten content of 30 wt.%. The predominant content of cobalt in nanoparticles relates to a lower boiling point of this metal and, as a consequence, its intensive evaporation from the surface of spheroidized granules in plasma. The presence of tungsten in nanoparticles may be caused by the removal of tungsten carbide nanoparticles from the surface of microparticles by the cobalt vapor during evaporation, as well as by the destruction of nanopowder microgranules during thermal decomposition of the organic binder with an active gas release.

The particle size distribution in the spheroidized powder after removal of nano- and submicron particles formed is shown in Figure 7. The disperse composition of the spheroidized powder is characterized by values D_{10} = 8 microns, D_{50} = 15 microns, and D_{90} = 28 microns. Compared with initial microgranules, the spherical particles are smaller (Figures 5 and 7), which may be related not only to the acquisition of a more compact spherical shape of particles but also to the destruction of microgranules in the plasma flow as a result of gas release at thermal decomposition of the organic binding used at granulation.

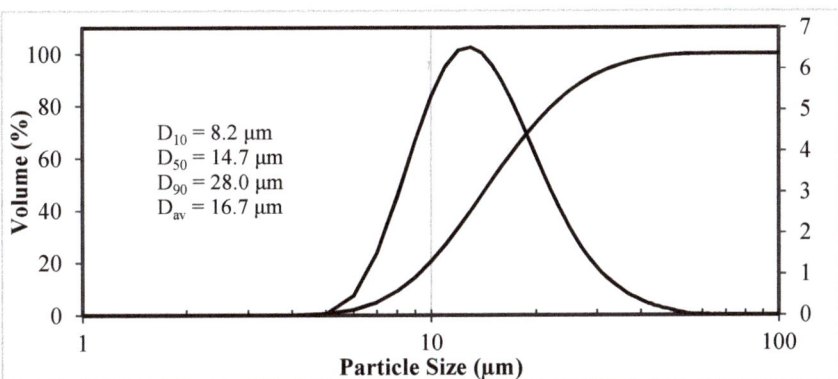

Figure 7. Differential and integral particle size distribution of W-C-Co micropowder after treatment in plasma and removal of nano- and submicron particles.

Thermal plasma jets generated by electric arc plasma torches are characterized by significant enthalpy and gas velocity gradients. In the plasma jet processing of polydisperse powders, the conditions of thermal interaction of particles of the processed material with the high-temperature gas are different. This fact predetermines possible differences in directions and rates of phase transformations occurring in individual particles, so that, under those conditions, particles in polydisperse powders may have different internal microstructures.

The change in phase composition in the process of nanopowder transformation into spherical microparticles is presented in Figure 8. The main phase in both nanopowder and spherical microparticles remains the tungsten carbide phase W_2C. When processing granules in a plasma flow the carbide phase $WC_{(1-x)}$ formation occurs and its content considerably increases with an increase in plasma stream enthalpy: by the results of XRD relative intensity of 100% reflections increases from 0.3 to 0.85. The content of tungsten monocarbide WC, obviously, also increases, as the ratio of relative intensity of 100% reflections WC and W_2C increases from 0.28 to 0.43. Increasing the enthalpy of the plasma jet during microgranules processing leads to an increase in the average mass temperature of the gas-dispersed stream and the microparticles present in it, which, in turn, may contribute to a rise in the content of the higher-temperature carbide phase $WC_{(1-x)}$.

In addition to the plasma enthalpy, there are several other factors that should significantly influence the phase composition of the W-C-Co spheroidized powder. These include the total carbon content of the nanopowder, the value of the specific surface area of the nanopowder, and the residence time of nanopowder granules in the high temperature region during plasma spheroidization. Within the framework of this work, the study of the influence of these factors was not carried out since the main task was to demonstrate the fundamental possibility of implementing the process of obtaining dense spherical microparticles of a composition based on tungsten carbides and cobalt with a submicron microstructure. A separate study is also required to determine the change in the phase composition and the structure of the resulting material in the L-PBF process using spheroidized micropowder of the W-C-Co system.

Spherical particles obtained in the experiments had predominantly dense microstructure and had no internal cavities, although some particles had small pores (Figures 6b and 9). Grain size of the particles in most cases was in the submicron range.

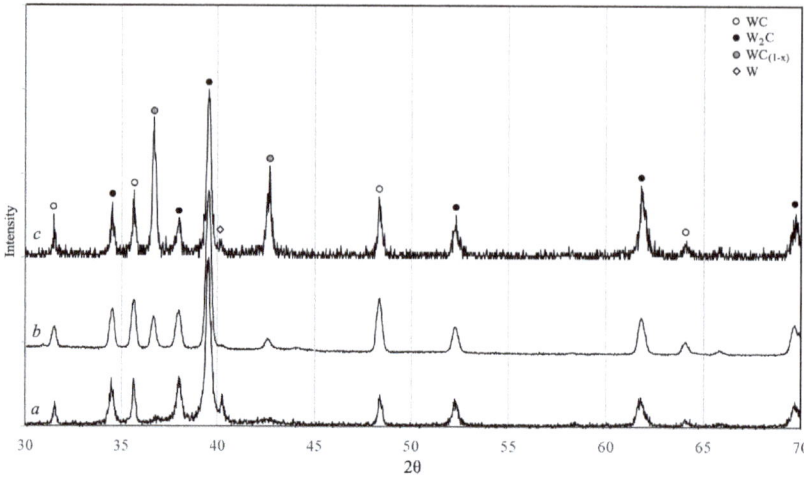

Figure 8. XRD patterns of initial W-C-Co nanopowder (**a**) and W-C-Co powder spheroidized at different plasma enthalpy: (**b**) 2.8 kW·h/m^3, (**c**) 4.8 kW·h/m^3.

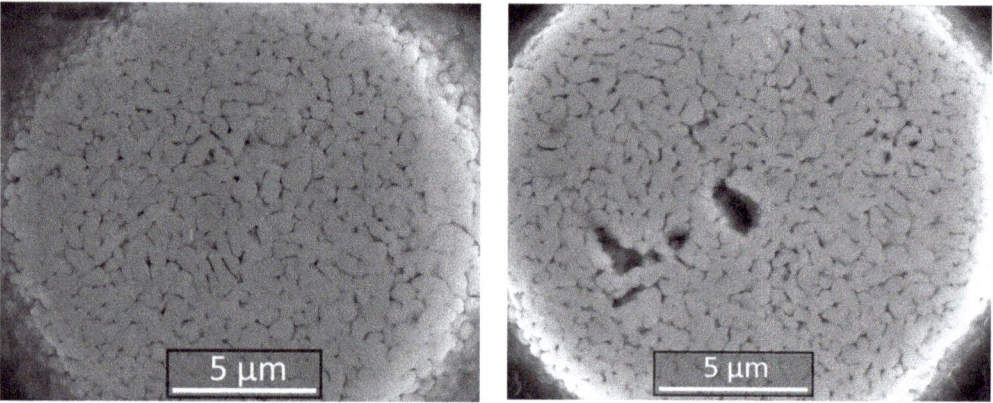

Figure 9. SEM images of W-C-Co spheroidized micropowder cross-section.

According to the EDS results of individual micropowder particles cross-sections with the most characteristic morphology revealed that the cobalt in them is uniformly distributed at submicron level in accordance with the structure of the particle (Figure 10).

As a result of elemental microanalysis of a cross-sectional area on ground spheroidal particles of W-C-Co composition (Figure 11), it is found that the amount of cobalt in particles of different size and structure varies in a wide range from 2.2 to 7.2 wt.% and averages about 5 wt.% for W+Co calculation (without carbon), which is approximately 4.7 wt.% for the W-Co system.

Intense evaporation of cobalt in the process of plasma spheroidization of granules led to a noticeable decrease in its total content in the obtained spherical powder. In initial microgranules the cobalt content (when analyzed by the XRF method) was 7.7 wt.%, and, after processing in a thermal plasma stream with enthalpy of 2.4 kW·h/m^3, the concentration of cobalt decreased to 4.6 wt.%. The treatment of granules in plasma at an enthalpy of 4.9 kW·h/m^3 reduced cobalt content in powder to 3.7 wt.% (Table 2).

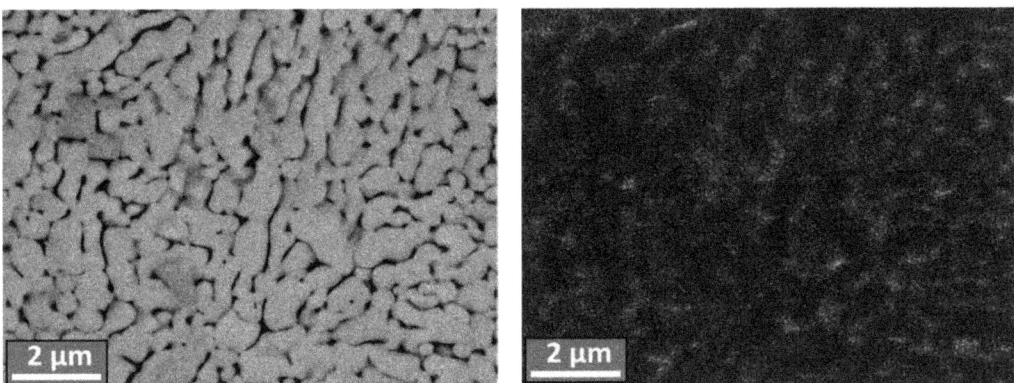

Figure 10. SEM image of W-C-Co spheroidized micropowder cross-section and map of Co distribution in this cross-section.

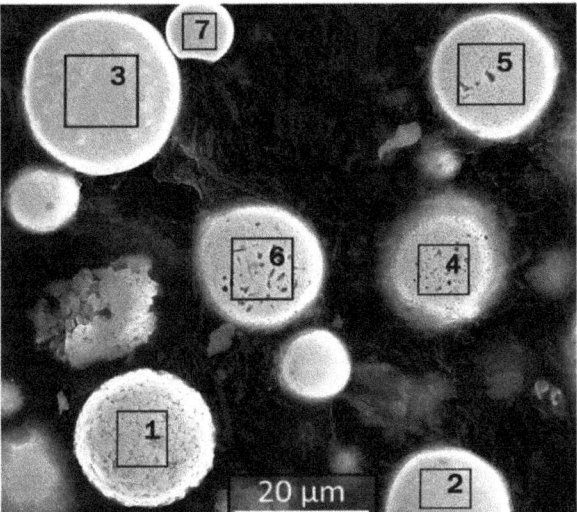

№	W, mass.%	Co, mass.%
1	92.8	7.2
2	97.3	2.7
3	93.9	6.1
4	94.7	5.3
5	93.6	6.4
6	94.0	6.0
7	97.8	2.2
Av.	94.9	5.1

Figure 11. SEM image of W-C-Co micropowder cross-section and the results of EDS elemental microanalysis showing the Co and W content in the powder particles with different structures.

Table 2. Composition of powders obtained.

	Nanopowder	Microgranules	Spheroidized Powder Produced at Different Enthalpy	
			2.8 kW·h/m^3	4.9 kW·h/m^3
Carbon, wt.%	4.7	6.4	4.7	3.9
Cobalt, wt.%	7.7	7.0	4.6	3.7
Oxygen, wt.%	0.5	1.0	0.05	0.03

When processing microgranules in a plasma flow, the carbon content in them also decreases. If the initial carbon content in granules before plasma treatment was 6.4 wt.%, after treatment in the plasma flow with enthalpy 2.4 kW·h/m^3, the carbon content decreased to 4.7 wt.% and for enthalpy 4.9 kW·h/m^3 to 3.9 wt.% (Table 2). Carbon in microgranules is present in the form of tungsten carbide phases, in a free state, and in PVB, which was used as a binder in the production of microgranules. The synthesized nanopowder had a carbon content of 4.7 wt.%, which suggests that when the microgranules were treated in a plasma flow, the carbon carryover was mainly due to the formation of gaseous carbon compounds during pyrolysis of PVB. Thus, the carbon introduced into the microgranules by the binder should not be accounted for in the carbon balance, which is involved in the chemical transformation of tungsten carbide phases during treatment/spheroidization of nanopowder microgranules in a plasma stream.

The oxygen content in the spheroidized powder was at the level of 0.03–0.05 wt.% and was determined by the intensity of initial granules interaction with plasma flow and its enthalpy level. The radical decrease in oxygen in powder from 1.0 wt.% in granules to 0.03 wt.% in spheroidized powder is determined first of all by the reductive chemical reactions of hydrogen with oxygen-containing granules components. The contribution of reactions in which the carbon of the W-C-Co composition and the carbon-containing products of thermal decomposition of the organic binder cannot be excluded.

4. Conclusions

The performed set of experimental studies has shown the principal possibility of obtaining dense spherical microparticles based on the composition of tungsten carbides and cobalt having a submicron structure by the consecutive use of plasma-chemical synthesis of a W-C-Co system nanopowder, spray drying of suspension based on a W-C-Co system nanopowder with obtaining nanopowder microgranules and spheroidization of microgranules in a thermal plasma flow.

It is found that the treatment of nanopowder microgranules in a thermal plasma stream leads to a change in their chemical composition: a reduction of carbon, oxygen, and cobalt, and the formation of nano- and submicron particles. To eliminate these negative effects, it is necessary to carry out further research aimed at investigating the development of methods to control the structure and chemical and phase composition of the obtained spherical microparticles of the tungsten carbide-cobalt system at the various stages in the process of their production in order to optimize their properties for the use of the micropowders in the manufacture of hard metal products by methods of additive technologies.

The study was supported by a grant from the Russian Science Foundation (project No. 19–73-00275).

Author Contributions: Conceptualization, A.S.; Data curation, A.A. and A.D.; Formal analysis, A.S., N.A. and M.S.; Investigation, A.A., A.F., M.S. and Y.K.; Methodology, A.A., A.F. and Y.K.; Resources, A.A., A.F. and Y.K.; Visualization, A.D.; Writing—original draft, N.A. and A.A.; Writing—review & editing, A.S., N.A and A.A. All authors have read and agreed to the published version of the manuscript.

Funding: This research was funded by Russian Science Foundation, grant number 19-73-00275.

Institutional Review Board Statement: Not applicable.

Informed Consent Statement: Not applicable.

Data Availability Statement: Not applicable.

Conflicts of Interest: The authors declare no conflict of interest.

References

1. Falkovskiy, V.A.; Klyachko, L.I. *Tverdye Splavi*; Ruda i Metally: Moscow, Russia, 2005. (In Russian)
2. Kurlov, A.S.; Gusev, A.I. *Tungsten Carbides: Structure, Properties and Application in Hardmetals*; Springer Series in Materials Science 184; Springer: Berlin/Heidelberg, Germany, 2016; p. 256.
3. Konyashin, I. Cemented carbides for mining, construction and wear parts. In *Comprehensive Hard Materials*; Mari, D., Miguel, L., Nebel, C.E., Eds.; Elsevier: Amsterdam, The Netherlands, 2014; pp. 425–451.
4. Yang, Y.; Zhang, C.; Wang, D.; Nie, L.; Wellmann, D.; Tian, Y. Additive manufacturing of WC-Co hardmetals: A review. *Int. J. Adv. Manuf. Technol.* **2020**, *108*, 1653–1673. [CrossRef]
5. Qian, M. Metal powder for additive manufacturing. *JOM* **2015**, *67*, 536–537. [CrossRef]
6. Saheb, S.H.; Durgam, V.K.; Chandrashekhar, A. A review on metal powders in additive manufacturing. *AIP Conf. Proc.* **2020**, *2281*, 020018.
7. Petrunichev, V.A.; Mihalev, V.I. K razrabotke protsessa plazmennogo polucheniya sfericheskih poroshkov iz tugoplavkih materialov. Izvestiya AN SSSR. *Metalli* **1966**, *6*, 154–158. (In Russian)
8. Rykalin, N.N.; Petrunichev, V.A.; Sorokin, L.M. Poluchenie sfericheskih tonkodispersnih poroshkov v nizkotemperaturnoi plazme. In *Plazmennie Protsessi v Metallurgii i Tehnologii Neorganicheskih Materialov*; Nauka: Moscow, Russia, 1973; pp. 220–230. (In Russian)
9. Boulos, M. Plasma power can make better powders. *Met. Powder Rep.* **2004**, *59*, 16–21. [CrossRef]
10. Vert, R.; Pontone, R.; Dolbec, R.; Dionne, L.; Boulos, M.I. Induction plasma technology applied to powder manufacturing: Example of titanium-based materials. In Proceedings of the 22nd International Symposium on Plasma Chemistry, Antwerp, Belgium, 5–10 July 2015.
11. Ji, L.; Wang, C.; Wu, W.; Tan, C.; Wang, G.; Duan, X.M. Spheroidization by plasma processing and characterization of stainless steel powder for 3d printing. *Metall. Mat. Trans. A* **2017**, *48*, 4831–4841. [CrossRef]
12. Kotlyarov, V.I.; Beshkarev, V.T.; Kartsev, V.E.; Ivanov, V.V.; Gasanov, A.A.; Yuzhakova, E.A.; Samokhin, A.V.; Fadeev, A.A.; Alekseev, N.V.; Sinayskiy, M.A.; et al. Production of spherical powders on the basis of group IV metals for additive manufacturing. *Inorg. Mater. Appl. Res.* **2017**, *8*, 452–458. [CrossRef]
13. Samokhin, A.; Alekseev, N.; Sinayskiy, M.; Astashov, A.; Kirpichev, D.; Fadeev, A.; Tsvetkov, Y.; Kolesnikov, A. Nanopowders Production and Micron-Sized Powders Spheroidization in DC Plasma Reactors. In *Powder Technology*; Cavalheiro, A.A., Ed.; IntechOpen: London, UK, 2018.
14. Samokhin, A.V.; Fadeev, A.A.; Alekseev, N.V.; Sinaysky, M.A.; Sufiyarov, V.S.; Borisov, E.V.; Korznikov, O.V.; Fedina, T.V.; Vodovozova, G.S.; Baryshkov, S.V. Spheroidization of Fe-based powders in plasma jet of DC arc plasma torch and application of these powders in selective laser melting. *Inorg. Mater. Appl. Res.* **2020**, *11*, 579–585. [CrossRef]
15. Itagaki, H.; Yachi, T.; Ogiso, H.; Sato, H.; Yamashita, Y.; Yasuoka, J.; Funada, Y. DC arc plasma treatment for defect reduction in WC-Co granulated powder. *Metals* **2020**, *10*, 975. [CrossRef]
16. Yan, Z.; Xiao, M.; Mao, X.; Khanlari, K.; Shi, Q.; Liu, X. Fabrication of spherical WC-Co powders by radio frequency inductively coupled plasma and a consequent heat treatment. *Powder Technol.* **2021**, *385*, 160–169. [CrossRef]
17. Fang, Z.Z.; Wang, X.; Ryu, T.; Hwang, K.S.; Sohn, H.Y. Synthesis, sintering, and mechanical properties of nanocrystalline cemented tungsten carbide—A review. *Int. J. Refract. Met. Hard Mater.* **2009**, *27*, 288–299. [CrossRef]
18. Lisovskii, A.F. On the development of nanostructured WC-Co hard alloys. *J. Superhard Mater.* **2010**, *32*, 389–395. [CrossRef]
19. GOST 19440-94. *Metallic Powders. Determination of Apparent Density*; IPC Publishing House of Standards: Moscow, Russia, 1996.
20. ISO 3923-1:2018. *Metallic Powders—Determination of Apparent Density—Part 1: Funnel Method*; Technical Committee ISO: Geneva, Switzerland, 2018.
21. ISO 3923-2. *Metallic Powders—Determination of Apparent Density—Part 2: Scott Volumeter Method*; Technical Committee ISO: Geneva, Switzerland, 1981.
22. Krasovskii, P.V.; Malinovskaya, O.S.; Samokhin, A.V.; Blagoveshchenskiy, Y.V.; Kazakov, V.A.; Ashmarin, A.A. XPS study of surface chemistry of tungsten carbides nanopowders produced through DC thermal plasma/hydrogen annealing process. *Appl. Surf. Sci.* **2015**, *339*, 46–54. [CrossRef]

Article

Fabrication of Silicon Carbide Fiber-Reinforced Silicon Carbide Matrix Composites Using Binder Jetting Additive Manufacturing from Irregularly-Shaped and Spherical Powders

Igor Polozov [1,*], Nikolay Razumov [1], Dmitriy Masaylo [1], Alexey Silin [1], Yuliya Lebedeva [2] and Anatoly Popovich [1]

1. Peter the Great St. Petersburg Polytechnic University, Polytechnicheskaya, 29, 195251 St. Petersburg, Russia; n.razumov@onti.spbstu.ru (N.R.); dmasaylo@gmail.com (D.M.); silin8888@mail.ru (A.S.); director@immet.spbstu.ru (A.P.)
2. Federal State Unitary Enterprise "All-Russian Scientific Research Institute of Aviation Materials" State Research Center of the Russian Federation, 17 Radio str., 105005 Moscow, Russia; yulia.ananieva@gmail.com
* Correspondence: polozov_ia@spbstu.ru

Received: 11 March 2020; Accepted: 7 April 2020; Published: 9 April 2020

Abstract: In this paper, silicon carbide fiber-reinforced silicon carbide (SiC_f/SiC) composites were fabricated using binder jetting additive manufacturing followed by polymer infiltration and pyrolysis. Spherical SiC powders were produced using milling, spray drying, and thermal plasma treatment, and were characterized using SEM and XRD methods. Irregularly shaped and spherical SiC powders were used to obtain SiC_f/SiC blends for the application in binder jetting. The effect of SiC powder shape on densification behavior, microstructure, and mechanical properties of binder jetted SiC_f/SiC composites was evaluated. The highest density of 2.52 g/cm^3 was obtained after six polymer infiltration and pyrolysis cycles. The microstructure and mechanical properties of the fabricated SiC_f/SiC composites were characterized. Using the spherical SiC powder resulted in higher fracture toughness and hardness, but lower flexural strength compared to the irregularly shaped powder. It was shown that it is feasible to fabricate dense SiC_f/SiC composites using binder jetting followed by polymer infiltration and pyrolysis.

Keywords: additive manufacturing; binder jetting; silicon carbide; spray drying; pyrolysis

1. Introduction

Silicon carbide fiber-reinforced silicon carbide (SiC_f/SiC) ceramic matrix composites (CMC) are considered to be promising materials for advanced applications in aerospace engines, gas turbines, and nuclear reactors [1]. Several methods have been used to make SiC_f/SiC particulate-based composites with pressure-assisted methods such as hot pressing, as well as pressed preforms that can be post-processed with sintering, liquid silicon infiltration (LSI), chemical vapor infiltration (CVI), and precursor impregnation and pyrolysis (PIP) [2–6]. The main limitation of these methods is the inability to produce parts with complex geometries. Machining of ceramics is an expensive and labor-consuming process; thus, achieving net shaped parts is important for saving costs. Often, up to 80% of the cost of manufacturing of ceramic parts may reflect the mechanical tooling cost [7].

Additive manufacturing (AM) offers a possibility to produce complex structures from various materials [8,9]. At the moment, the majority of AM processes, excluding those used for manufacturing parts of polymers, use a laser or an electron beam as the energy source, and metal powder or wire as the feedstock material. Such an approach limits the number of materials used in AM, in particular, for intermetallic and ceramic materials. At the same time, demand for ceramic parts for the aerospace and energy industries has been constantly increasing [10].

Binder jetting technology, which utilizes powders as feedstock material, is one of the methods for AM of ceramic components, as Gonzalez et al. [11] and Du et al. [12] showed using alumina powders. Binder jetting involves depositing a binder on a powder layer and curing the binder to obtain a green part [13]. One of the major drawbacks of this technology is the high porosity and poor mechanical properties of the produced green parts [14,15]. This results from low powder packing density in an applied powder layer as well as the use of a binder to build a part, which leads to a weak connection between powder particles as opposed to other AM process.

Stereolithography and robocasting are other AM methods that can be used to fabricate ceramic parts [16,17]. These methods require the preparation of colloidal gels and photocuring slurries, which might be time-consuming. Moreover, SiC exhibits a high refractive index and light absorption in the ultraviolet wavelength, which makes it challenging for the stereolithography [18].

Even though the parts obtained by binder jetting are characterized by a high level of porosity, it is still possible to achieve a high density of ceramic parts after subsequent sintering or infiltration. For example, Gonzalez et al. [11] achieved 96% of the theoretical density for aluminum oxide parts after sintering at 1600 °C for 16 h. Fielding et al. [19] achieved 95% density after sintering by adding ZnO and SiO_2 to $(Ca_3(PO_4)_2)$ powder. Melcher et al. [20] infiltrated alumina green samples having 36% density using copper to fabricate dense parts. A similar process was used to infiltrate alumina samples with glass by Zhang et al. [21]. Another example of such an approach is the production of reactive-bonded ceramic based on SiC; Fu et al. [22] used a mixture of silicon, silicon carbide, and dextrin to obtain samples by binder jetting followed by pyrolysis and liquid silicon infiltration, which resulted in reactive-bonded silicon carbide parts. However, such an approach limits the maximum working temperature, since silicon has a melting point of 1414 °C. Moon et al. [23] fabricated a carbon green-part by binder jetting and then infiltrated it with silicon, which resulted in the formation of a reactive-bonded SiC-based composite material. Monolithic SiC samples with >90% of theoretical density have been fabricated by Terrani et al., by combining binder jetting and chemical vapor infiltration processes as shown in [24]. Fleisher et. al. [25] used binder jetting followed by phenolic resin binder impregnation and capillary liquid silicon infiltration to successfully fabricate complex-shaped SiC parts; however, residual silicon would limit the maximum operating temperature of such parts.

Even though the binder jetting process has been applied to fabricate parts from various ceramic materials, the studies devoted to SiC_f/SiC fabrication are limited. In a previous paper [26], SiC_f/SiC composites were fabricated using the binder jetting process followed by PIP. However, only irregular-shaped SiC powder particles were used. Spherical powders are considered to be preferable in AM processes, including binder jetting technology, since they provide better flowability, packing density, and sinterability compared to irregularly shaped particles [27,28]. Thus, it is important to investigate the effect of powder shape and morphology on the properties of binder jetted SiC_f/SiC CMCs.

The main objective of this paper was to investigate the fabrication process of SiC_f/SiC CMCs by binder jetting AM technology using powder compositions based on both irregular and spherical SiC particles with SiC fibers. The feasibility of spray drying and plasma treatment processes to fabricate spherical silicon carbide powder for AM applications was investigated. The effect of SiC powder shape on densification behavior, microstructure, and mechanical properties of binder jetted SiC_f/SiC composites was evaluated.

2. Materials and Methods

Silicon carbide F320 grit powders (d_{50} = 38.3 µm) with irregular (non-spherical) shape and Si-TUFF™ silicon carbide fibers with 100–300 µm of length and 7 µm of average diameter were used in the binder jetting process. The SiC fibers and SiC powders were blended in a tumbler mixture for 12 h to obtain a composite mixture.

Prior to spray drying, SiC F320 grit powders were milled in a jet-type mill Netzsch CGS10 using the following process parameters: gas pressure = 7 atm., classifier's rotation speed = 17,000 rpm, powder feeding rate = 50 g/min. The selection of the milling parameters was based on preliminary experiments. The milled powder had the following particle size distribution: d_{10} = 0.7 µm, d_{50} = 1.4 µm, d_{90} = 2.5 µm.

Silicon metal powder (99.9%) up to 5 µm in size was mixed with the milled SiC powder prior to the spray drying to act as a binder during the subsequent plasma spheroidization.

The spray drying process of the 95 SiC-5 Si (wt.%) powder blend was carried out using a LPG-5 spray dryer (Changzhou Yibu Drying Equipment Co. Ltd, Jiangsu, China). The spray dryer operating principle is the following. Filtered heated air is uniformly fed as a spiral downflow to a drying chamber through an air dispenser. The liquid material is sprayed inside the drying chamber through a centrifugal disk pulverizer. Upon contact with the hot air, the fine droplets of the pulverized liquid are dried and form separate fine particles falling on the chamber floor. A 10% water solution of polyvinyl alcohol (PVA) was used as a binder with a 1:1 weight ratio of the powder to the binder. The air temperature during the spray drying was 170–180 °C. The pulverizing disk rotating frequency was varied between 15–50 Hz.

Plasma jet treatment experiments were carried out using TekSphero 15 plasma spheroidization equipment manufactured by Tekna Plasma Systems Inc. (Sherbrooke, Québec, Canada). The preliminary experiments were carried out to choose the following plasma treatment process parameters to obtain SiC spherical particles: plasma torch power = 15 kW, plasma gas (argon hydrogen mix) flow rate = 2.4 m^3/h, powder feeding rate = 10–70 g/min.

The ExOne Innovent binder jetting printer and ExOne solvent binder were used to build samples of 10 × 7 × 70 mm^3 in size. The following printing parameters were used: 4–7 mm/s recoat speed, 20 mm/s roller speed, 120 rpm roller rotation speed, 30 s dry time at 60 °C drying temperature, 100 µm layer thickness, and 60% binder saturation. The green parts were cured at 190 °C for 3 h. The binder jetting process parameters were preliminary optimized to achieve a steady fabrication process of the samples with a desired geometry.

Polycarbosilane with StarPCS^{TM} SMP-10 trade name manufactured by Starfire Systems (Schenectady, NY, USA) was used for infiltration of the green parts. The infiltration process was carried out in a vacuum chamber for 1 h by vacuuming the chamber with the samples and then introducing SMP-10 into the chamber so that it fully covered the samples.

After the SMP-10 infiltration, the samples were subjected to pyrolysis in a heating furnace at 1000 °C for 1 h with an argon flow. The heating rate was 10 °C/min. An intermediate dwelling of the samples at 500 °C for 30 min was done to remove the gases forming during the heating of SMP-10. These parameters were chosen based on the data showed by [26] for SMP-10 pyrolysis used for binder jetted SiC samples, as well as for SMP-10 pyrolysis of electrophoretic deposited SiC-samples [29].

A summary of the types of SiC powders and treatments used in the current study is presented in Table 1.

Table 1. Summary of the types of SiC powders and treatments used in the study.

Type of Powder	Powder Treatment	Fibers Used	Fabrication of Samples	Post-Treatment of Samples
Irregular SiC (d_{50} = 38.3 µm)	None (1) Milling; (2) Mixing with 5 wt.% of Si powder; (3) Spray drying; (4) Plasma treatment	30% vol. of SiC fibers mixed with the SiC powder	Binder jetting	SMP-10 infiltration and pyrolysis

Three-point flexural tests were carried out according to ISO 17138:2014 "Fine ceramics (advanced ceramics, advanced technical ceramics)—Mechanical properties of ceramic composites at room temperature—Determination of flexural strength" with the samples of 7 × 10 × 70 mm^3 size.

The hardness H_v and fracture toughness K_{1C} of the fabricated ceramic samples were measured using a Buehler VH1150 testing machine (Buehler, Lake Bluff, IL, USA) with a 1000-g load and 10 s of dwell time. The hardness values were calculated according to Formula (1):

$$H_v = 1.854 \cdot \frac{g \cdot F}{(2a)^2} \cdot 10^{-3}, \tag{1}$$

where F is the load, H; a is the length of a semi-diagonal of the indentation, µm; l is the crack length, µm; g is the acceleration of gravity (9.8 m/s^2).

Fracture toughness was calculated according to the formula for the Palmqvist crack model (2):

$$K_{Ic} = 0.048 \left(\frac{l}{a}\right)^{-1/2} \left(\frac{H_v}{E\Phi}\right)^{-2/5} \left(\frac{H_v a^{1/2}}{\Phi}\right), \qquad (2)$$

where E is Young's module (taken as 360 GPa in this study), Φ is 3; H_v is the hardness; a is the length of a semi-diagonal of the indentation, μm; l is the crack length, μm.

The fracture toughness measurement method with a Vickers indenter was suggested by Evans and Wilshaw [30] and then further extended by Niihara [31]. The equations for calculating the fracture toughness using the Vickers indenter method are based on a semi-empirical calculation between the indentation load and the length of the cracks coming from the corners of the indent.

The phase composition was analyzed with a Bruker D8 Advance X-ray diffraction (XRD) (Bruker, Billerica, MA, USA) meter using Cu-Kα (λ = 1.5418 Å) irradiation. The microstructure investigation and powder morphology studies were carried out using TESCAN Mira 3 LMU scanning electron microscope (SEM) (Tescan, Brno, Czechia).

The fabricated samples were measured using a digital caliper with a resolution of 0.01 mm and the results were compared to the CAD-data.

3. Results

3.1. Powder Characterization

Figure 1 shows the initial SiC powder particles and the SiC fibers. The SiC powder features an irregular particle shape with smooth faces and has the following particle size distribution: d_{10} = 21.8 μm, d_{50} = 38.3 μm, d_{90} = 63.2 μm. The apparent density of the SiC powder is 1.37 g/cm³. The SiC fibers (Figure 1b) have a length of about 100–300 μm with a diameter of 7–10 μm.

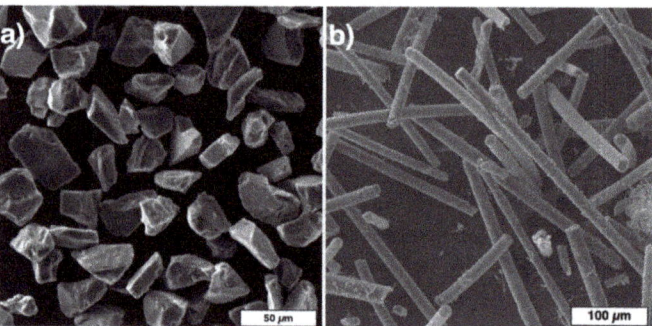

Figure 1. Initial silicon carbide (SiC) powder particles (**a**) and SiC fibers (**b**).

According to the Si-C phase diagram [32], SiC melts incongruently at 2830 °C and starts to decompose at about 1800 °C [33]. During its treatment in a thermal plasma jet, the initial powder partially sublimates followed by condensation of fine round particles with a size below 1 μm (Figure 2a). Epitaxial growth of columnar crystals also takes place at the surface of coarse particles (Figure 2b). Thus, a three-step approach has been suggested to obtain spherical SiC powder particles, which involves obtaining particles consisting of SiC, Si particles, and a binder using the spray drying process and then subjecting these particles to the plasma treatment.

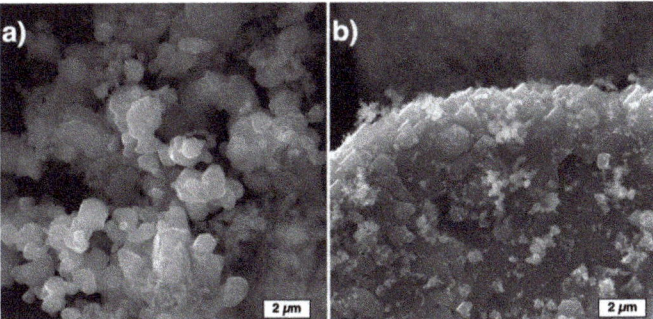

Figure 2. SiC F320 particles after treatment in a thermal plasma jet showing (**a**) condensated particles and (**b**) a particle's surface morphology.

During the spray drying process, agglomerated spherical particles with the size 10–80 μm (Figure 3) were fabricated. The agglomerated particle surface consists of evenly distributed fine particles (Figure 3b). Two types of particles can be distinguished by their size: coarse particles (1–4 μm) and fine particles of submicron size. The fine particles fill the voids between the coarse particles.

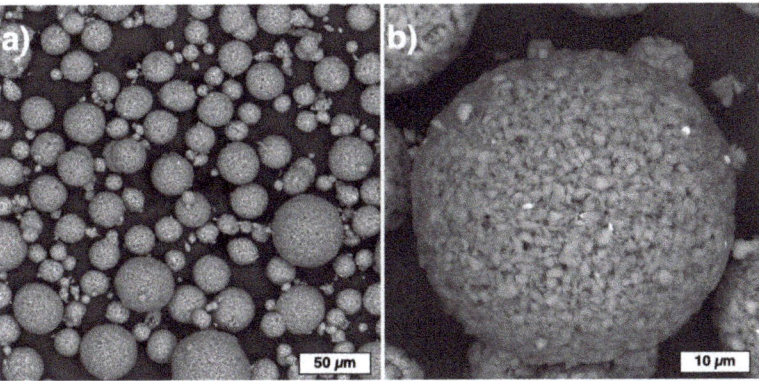

Figure 3. Powder particles obtained by spray drying of 95% SiC-5% Si slurry with polyvinyl alcohol (PVA) as the binder: (**a**) general view and (**b**) a particle's surface morphology.

Figure 4 shows the spherical powder particles obtained by the treatment of spray dried powders in a thermal argon hydrogen plasma jet. The obtained powder particles do not have internal voids and feature a homogeneous chemical distribution as confirmed by SEM investigations with back-scattered electrons. Local sintering of SiC particles occurs during the plasma treatment, which results in necking between the adjacent particles. During the plasma treatment, silicon melted and binded the SiC particles, while the binder dissociated with the formation of carbon. Carbon reacted with silicon and formed a reactive-bonded SiC. Similarly, the formation of β-SiC was reported in [34] when SiO and C powders were treated in a radiofrequency thermal plasma. The particle size distribution of the obtained powder after the plasma treatment was the following: d_{10} = 24.4 μm, d_{50} = 49.5 μm, d_{90} = 89.9 μm. The amount of powder particles below 10 μm was about 1.8% vol. The apparent density of the powder was 0.69 g/cm^3, which is lower compared to the initial F320 powder due to finer SiC particles forming the spherical particles.

Figure 4. SiC powder particles after spray drying and plasma treatment: (**a**) general view and (**b**) a particle's surface morphology.

As expected, during the plasma treatment of the spray dried powder, pure silicon was melted and bonded the SiC particles (Figure 5). The presence of the silicon was confirmed by the X-ray analysis illustrated below. At the same time, silicon partially reacted with the carbon formed from PVA dissociation, which resulted in the formation of the secondary SiC.

Figure 5. SiC powder particles after spray drying and plasma treatment in secondary electrons (**a**) and backscattered electrons (**b**).

SiC powders mixed with 30% vol. of SiC fibers were used for application in the binder jetting process. Both irregular and spherical SiC powders were used to prepare the separate composite blends as shown in Figure 6.

Figure 6. Composite blends prepared from SiC powders with irregular (**a**) and spherical (**b**) particles and 30% vol. of SiC fibers.

3.2. Densification of the SiCf/SiC Samples During Infiltration and Pyrolysis

The binder jetting process was used to prepare the samples from the composite blends. The samples fabricated using irregular SiC particles had a bulk density of 1.34–1.39 g/cm^3, while the samples obtained using spherical SiC particles had a bulk density of 0.72 g/cm^3, as measured using the Archimedes method. The obtained density values correspond to 56–58% of porosity for the samples with irregular-shaped particles and 76% for the samples with spherical particles. High porosity values for the samples with spherical particles are due to the presence of submicron pores in the initial spherical SiC particles after the plasma treatment.

The fabricated green samples were subjected to several cycles of SMP-10 infiltration and pyrolysis. SMP-10 is known to have a good wetting ability that allows infiltrating very small pores [29]. During the pyrolysis process, SMP-10 polycarbosilane in the sample's pores forms into secondary silicon carbide. Hence, the porosity of the samples decreases. Figure 7 shows the changes in density and porosity of the samples after various numbers of infiltration and pyrolysis cycles. After the first cycle, the density of SiC$_f$/SiC samples, fabricated using irregular particles, increases from 1.36 g/cm^3 to 1.71 g/cm^3. The porosity decreases from 56% to 46%. For the samples fabricated using spherical particles, the density increases from 0.72 g/cm^3 to 1.81 g/cm^3, decreasing the porosity from 76% to 46%. The ceramic yield of the liquid polycarbosilane is approximately 60–70% [35,36]. Hence, a large volume fraction of pores still remains after the pyrolysis. While increasing the number of infiltration and pyrolysis cycles leads to higher density, the degree of impact decreases due to filling up the pores close to the surface of the samples with the secondary SiC and preventing the SMP-10 to further infiltrate the samples. Similarly, Halbig et al. [26] showed that the highest increase in density for binder jetted SiC samples occurred in the first two infiltration steps. After the sixth cycle, the density and porosity values for the sample fabricated from irregular particles reached 2.52 g/cm^3 and 20%, respectively, while for the sample fabricated from spherical particles the density and porosity were 2.21 g/cm^3 and 24%, respectively. A similar trend was reported in [36], when SiC$_f$/SiC composites were fabricated by PIP of continuous SiC fiber preforms, however lower densities were achieved after the same number of PIP cycles.

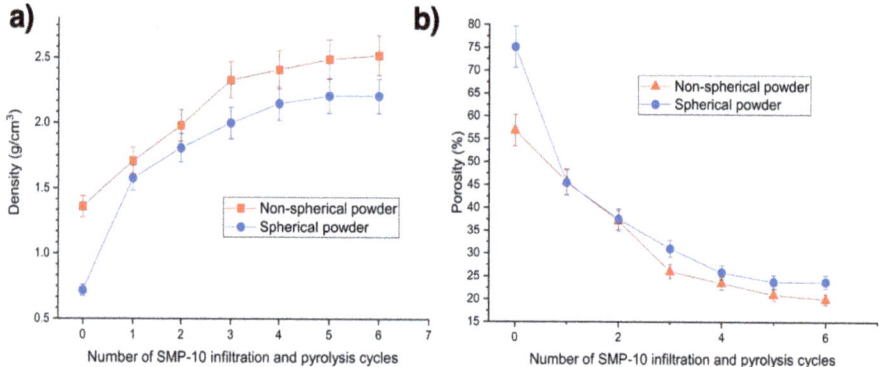

Figure 7. Change in density (**a**) and porosity (**b**) of the SiCf/SiC samples with the number of infiltration and pyrolysis cycles.

3.3. Microstructure Characterization

Figure 8 shows SEM images of the microstructure of the SiC$_f$/SiC samples fabricated from spherical particles after six cycles of infiltration and pyrolysis (Figure 8a), and the fracture surface of the sample (Figure 8b). The microstructure features spherical SiC particles and separate SiC fibers with a small amount of Si particles. After several infiltrations and pyrolysis cycles, SMP-10 polycarbosilane formed a SiC phase, which resulted in partially closing the internal pores. The majority of the residual pores is located in coarse spherical SiC particles remaining from the initial particles. These pores are filled up during the first SMP-10 infiltration, which prevents further infiltration of these pores. The fracture surface of the samples features the initial separate powder particles of SiC and SiC fibers that are bonded by secondary SiC formed from SMP-10. There are also interconnected pores that might act as stress concentrators during the load and initiate crack formation.

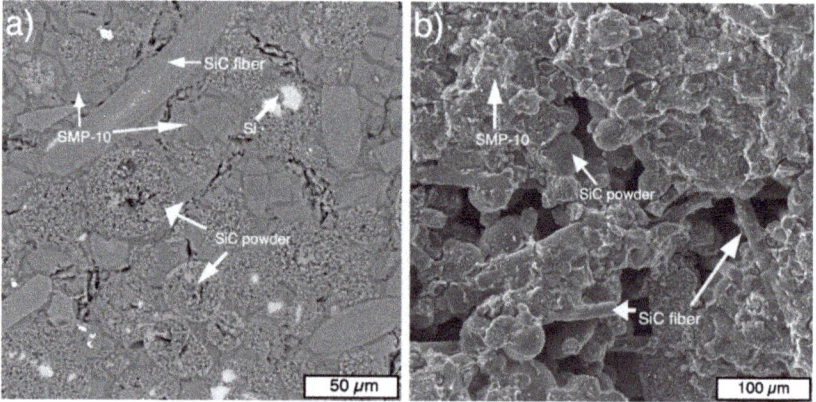

Figure 8. SEM images of the microstructure (**a**) and fracture surface (**b**) of the SiCf/SiC sample fabricated from spherical particles after six infiltration and pyrolysis cycles.

In the case of the sample fabricated from irregular SiC powders, the microstructure features SiC powder particles with the initial shape and separate SiC fibers. The secondary SiC also formed between powder particles from SMP-10 after pyrolysis (Figure 9a). The pores are also present in the sample, but visually, the sample obtained from irregular particles has a denser structure compared to the sample obtained from spherical particles, which was confirmed by Archimedes measurements. The fracture surface features the initial irregular SiC particles bonded by the secondary SiC, as well as SiC fibers (Figure 9b).

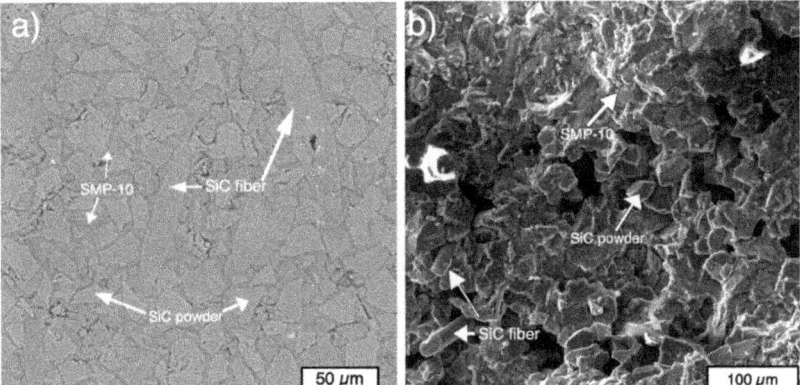

Figure 9. SEM-images of the microstructure (**a**) and fracture surface (**b**) of the SiCf/SiC sample fabricated from irregular particles after six infiltration and pyrolysis cycles.

Figure 10 shows the XRD results for the initial powders and the fabricated samples. The XRD pattern of the initial powder is characterized by the presence of two SiC modifications: α-SiC peaks with Moissanite-6H and Moissanite-15R crystal structure with an approximate content of 95.6 and 4.4%, respectively. The initial α-SiC powder with Moissanite-6H crystal structure is characterized by the following unit cell parameters: a = 0.3081 nm and c = 1.5114 nm; and Moissanite-15R crystal structure has a = 0.3081 nm and c = 3.7782 nm. The samples 2–4 feature silicon with Fd-3m crystal structure in the amount of 4.4–4.8%. Peak broadening occurred for the powders after spray drying is associated with the coherent scattering region size decrease (down to 50.1 nm).

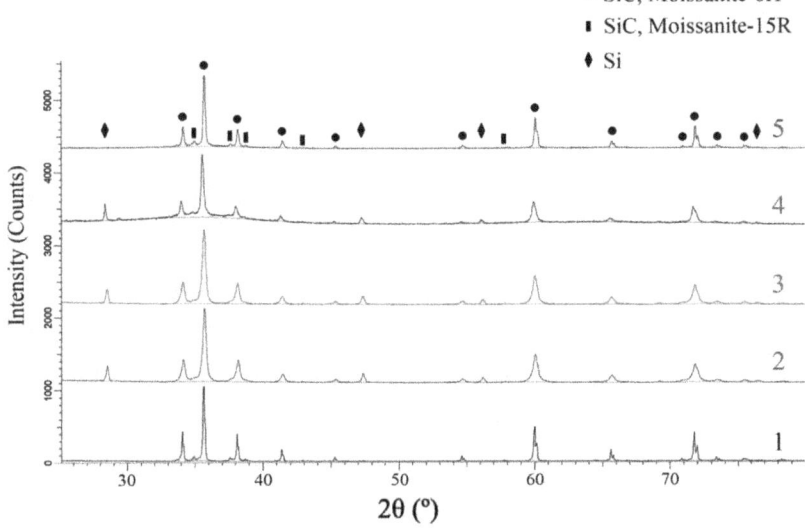

Figure 10. XRD results for the initial SiC powder (**1**), for the powder obtained by spray drying (**2**) and plasma treatment (**3**), and for the samples fabricated by binder jetting followed by SMP-10 infiltration and pyrolysis from spherical (**4**) and irregular (**5**) SiC powders.

The ratio of α-SiC content with Moissanite-6H and Moissanite-15R crystal structure in the samples fabricated by binder jetting followed by SMP-10 infiltration and pyrolysis remained the same as in the initial powders. The samples 4 and 5 have the following Moissanite-6H crystal unit cell parameters: a = 0.3080 nm and c = 1.5107 nm, and a = 0.3080 nm and c = 1.5112 nm, respectively, which are close to the reference values for SiC. The coherent scattering region size values are 59.3 nm (for sample 4) and 134 nm (for sample 5). The XRD pattern for sample 4 is characterized by the increased background noise at the peaks' bottom, suggesting the presence of a small amount of amorphous SiC phase. No phase transformations of silicon carbide were found for different samples.

3.4. Mechanical Properties

The three-point bending test results showed that the SiC$_f$/SiC samples fabricated from the irregular SiC particles have an average flexural strength of 118.7 MPa, while the samples fabricated from the spherical SiC particles have an average flexural strength of 62.7 MPa, as shown in Figure 11. Both materials demonstrate almost linear stress-displacement curves under the flexural load, indicating brittle fracture.

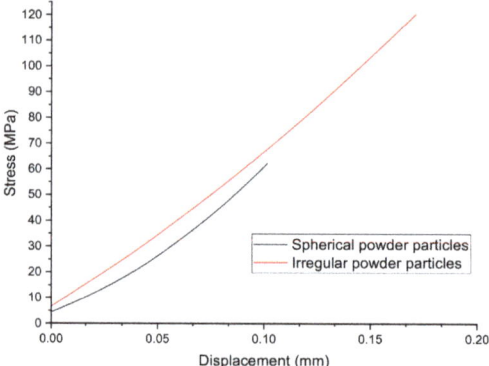

Figure 11. Typical curves for flexural stress-displacement data obtained from the samples produced using spherical and irregular SiC powder particles.

The lower flexural strength values for the samples obtained from spherical particles are the result of a higher pore volume fraction. At the same time, microhardness and fracture toughness are higher for these samples since the separate SiC particles are significantly finer in case of spherical particles, which inhibits crack growth. The measured fracture toughness (shown in Table 2) for the samples from spherical powders is comparable to the values achieved by Hayun et al. [37] for dense (3.18 g/cm^3) silicon carbide samples fabricated by spark plasma sintering; however, hardness and bending strength values are notably lower. The binder jetted composite SiC$_f$/SiC samples showed lower hardness and flexural strength than sintered dense (>98% of the theoretical density) α-SiC material (around 30 GPa hardness and 300 MPa four-point bending strength) [38]. However, the fracture toughness of the binder jetted composite fabricated from the spherical powder is approximately two times higher. The obtained flexural strength values are lower compared to the conventionally fabricated SiC$_f$/SiC composites. For example, tape-casted and hot-pressed SiC short-fiber-reinforced SiC composites demonstrated a flexural strength of about 370 MPa [39], but their density was higher, which benefited the mechanical properties. On the other hand, the fracture toughness of the hot-pressed composites was lower (3.23 MPa·m$^{1/2}$) compared to the binder jetted samples. Thus, a further investigation of porosity reduction for binder jetted silicon carbide samples as well as heat treatment effects is necessary to achieve enhanced mechanical properties of the binder jetted SiC$_f$/SiC composite. Silicon melt infiltration is one of the possible ways to reduce porosity and increase strength; however, the presence of pure silicon would limit a maximum operating temperature [40]. Another approach to increase the flexural strength of binder jetted SiC$_f$/SiC samples might be to increase the volume fraction of SiC fibers and deposit a protective coating on the fibers, which prevents degradation of the fibers and improves the mechanical properties of CMCs as reported in [41,42].

Table 2. Average values of microhardness and fracture toughness for SiC$_f$/SiC samples.

SiC Powder Particles	Hv. GPa	Fracture Toughness K1C, MPa·m$^{1/2}$
Irregular	11.6	3.58
Spherical	20.8	6.13

The presented approach to fabricate SiC$_f$/SiC composite parts using binder jetting AM process followed by polycarbosilane infiltration and pyrolysis was demonstrated using a turbine blade prototype model as shown in Figure 12a for as-fabricated green-part and in Figure 12b for the part after six cycles of infiltration and pyrolysis. The binder jetting process followed by several infiltration and pyrolysis cycles has been shown to be a promising method to fabricate SiC$_f$/SiC parts enabling complex geometry of parts as well as good dimensional accuracy (about 150–200 µm) and surface finish.

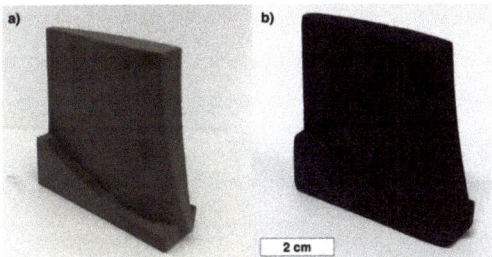

Figure 12. As-fabricated SiC$_f$/SiC turbine blade prototype (**a**) and the part after six cycles of infiltration and pyrolysis (**b**).

4. Conclusions

The present paper demonstrates the feasibility of binder jetting AM of SiC fiber-reinforced SiC composite material followed by polycarbosilane infiltration and pyrolysis. Both irregular and spherical SiC powder particles were used for fabrication of the samples by binder jetting.

The spherical SiC powder particles were obtained by milling the initial irregular SiC powder followed by spray drying and plasma jet treatment. It was demonstrated that the spray drying process followed by thermal plasma treatment can be used for the manufacturing of spherical SiC powder particles with a size of 10–80 µm.

The infiltration and pyrolysis of binder jetted SiC$_f$/SiC samples result in a density increase and porosity decrease due to the formation of a secondary SiC phase from SMP-10 polycarbosilane inside the pores. The highest density values (2.52 g/cm^3 and 2.21 g/cm^3 for the samples fabricated from irregular and spherical powders, respectively) were achieved after six infiltration and pyrolysis cycles. The difference in the final densities is associated with the presence of submicron pores in spherical powders, which inhibit the further filling of the pores with SMP-10.

The microstructure of the binder jetted samples consists of the initial SiC powder particles that maintained their shape and SiC fibers bonded with a secondary SiC phase.

The three-point bending tests showed that the SiC$_f$/SiC samples fabricated from the irregular SiC particles have an average flexural strength of 84.3 MPa, and the samples fabricated from the spherical SiC particles have an average flexural strength of 52.9 MPa. The lower flexural strength values for the samples obtained from spherical particles are the result of higher internal porosity. Microhardness and fracture toughness are higher for the samples fabricated from spherical particles, since the separate SiC particles are significantly finer in this case, which inhibits crack growth.

Author Contributions: Conceptualization, I.P., N.R. and D.M.; Funding acquisition, A.P.; Investigation, I.P. and A.S.; Methodology, I.P. and N.R.; Resources, Y.L.; Supervision, Y.L. and A.P.; Visualization, I.P.; Writing—Original draft, N.R.; Writing—Review & editing, D.M. All authors have read and agreed to the published version of the manuscript.

Funding: This research received no external funding.

Conflicts of Interest: The authors declare no conflict of interest. The funders had no role in the design of the study; in the collection, analyses, or interpretation of data; in the writing of the manuscript, or in the decision to publish the results.

References

1. Yoshida, K. Development of silicon carbide fiber-reinforced silicon carbide matrix composites with high performance based on interfacial and microstructure control. *J. Ceram. Soc. Jpn.* **2010**, *118*, 82–90. [CrossRef]
2. Yin, J.; Lee, S.-H.; Feng, L.; Zhu, Y.; Liu, X.; Huang, Z.; Kim, S.-Y.; Han, I.-S. The effects of SiC precursors on the microstructures and mechanical properties of SiCf/SiC composites prepared via polymer impregnation and pyrolysis process. *Ceram. Int.* **2015**, *41*, 4145–4153. [CrossRef]
3. Dong, S.; Katoh, Y.; Kohyama, A. Preparation of SiC/SiC Composites by Hot Pressing, Using Tyranno-SA Fiber as Reinforcement. *J. Am. Ceram. Soc.* **2003**, *86*, 26–32. [CrossRef]
4. Kotani, M.; Kohyama, A.; Okamura, K.; Inoue, T. Fabrication of high performance SiC/SiC composite by polymer impregnation and pyrolysis method. In *Ceramic Engineering and Science Proceedings*; American Ceramic Society: Westerville, OH, USA, 1999; pp. 309–316.
5. Kotani, M.; Kohyama, A.; Katoh, Y. Development of SiC/SiC composites by PIP in combination with RS. *J. Nucl. Mater.* **2001**, *289*, 37–41. [CrossRef]
6. Wilhelm, M.; Kornfeld, M.; Wruss, W. Development of SiC–Si composites with fine-grained SiC microstructures. *J. Eur. Ceram. Soc.* **1999**, *19*, 2155–2163. [CrossRef]
7. Klocke, F. Modern approaches for the production of ceramic components. *J. Eur. Ceram. Soc.* **1997**, *17*, 457–465. [CrossRef]
8. Mirzendehdel, A.M.; Suresh, K. Multi-Material topology optimization for additive manufacturing. In Proceedings of the ASME 35th Computers and Information in Engineering Conference, Boston, MA, USA, 2–5 August 2015; 2015; Volume 1A, p. V01AT02A020.
9. DebRoy, T.; Wei, H.L.; Zuback, J.S.; Mukherjee, T.; Elmer, J.W.; Milewski, J.O.; Beese, A.M.; Wilson-Heid, A.; De, A.; Zhang, W. Additive manufacturing of metallic components—Process, structure and properties. *Prog. Mater. Sci.* **2018**, *92*, 112–224. [CrossRef]
10. Eswara Prasad, N.; Kumar, A.; Subramanyam, J. Ceramic Matrix Composites (CMCs) for aerospace applications. In *Aerospace Materials and Material Technologies*; Springer: Singapore, 2017; pp. 371–389.
11. Gonzalez, J.A.; Mireles, J.; Lin, Y.; Wicker, R.B. Characterization of ceramic components fabricated using binder jetting additive manufacturing technology. *Ceram. Int.* **2016**, *42*, 10559–10564. [CrossRef]
12. Du, W.; Ren, X.; Ma, C.; Pei, Z. Ceramic binder jetting additive manufacturing: Particle coating for increasing powder sinterability and part strength. *Mater. Lett.* **2019**, *234*, 327–330. [CrossRef]
13. Snelling, D.A.; Williams, C.B.; Suchicital, C.T.A.; Druschitz, A.P. Binder jetting advanced ceramics for metal-ceramic composite structures. *Int. J. Adv. Manuf. Technol.* **2017**, *92*, 531–545. [CrossRef]
14. Tang, Y.; Zhou, Y.; Hoff, T.; Garon, M.; Zhao, Y.F. Elastic modulus of 316 stainless steel lattice structure fabricated via binder jetting process. *Mater. Sci. Technol.* **2016**, *32*, 648–656. [CrossRef]
15. Rabinskiy, L.; Ripetsky, A.; Sitnikov, S.; Solyaev, Y.; Kahramanov, R. Fabrication of porous silicon nitride ceramics using binder jetting technology. *IOP Conf. Ser. Mater. Sci. Eng.* **2016**, *140*, 012023. [CrossRef]
16. Travitzky, N.; Bonet, A.; Dermeik, B.; Fey, T.; Filbert-Demut, I.; Schlier, L.; Schlordt, T.; Greil, P. Additive manufacturing of ceramic-based materials. *Adv. Eng. Mater.* **2014**, *16*. [CrossRef]
17. Cai, K.; Román-Manso, B.; Smay, J.E.; Zhou, J.; Osendi, M.I.; Belmonte, M.; Miranzo, P. Geometrically Complex Silicon Carbide Structures Fabricated by Robocasting. *J. Am. Ceram. Soc.* **2012**, *95*, 2660–2666. [CrossRef]
18. de Hazan, Y.; Penner, D. SiC and SiOC ceramic articles produced by stereolithography of acrylate modified polycarbosilane systems. *J. Eur. Ceram. Soc.* **2017**, *37*, 5205–5212. [CrossRef]
19. Fielding, G.A.; Bandyopadhyay, A.; Bose, S. Effects of silica and zinc oxide doping on mechanical and biological properties of 3D printed tricalcium phosphate tissue engineering scaffolds. *Dent. Mater.* **2012**, *28*, 113–122. [CrossRef] [PubMed]
20. Melcher, R.; Martins, S.; Travitzky, N.; Greil, P. Fabrication of Al_2O_3-based composites by indirect 3D-printing. *Mater. Lett.* **2006**, *60*, 572–575. [CrossRef]
21. Zhang, W.; Melcher, R.; Travitzky, N.; Bordia, R.K.; Greil, P. Three-Dimensional Printing of Complex-Shaped Alumina/Glass Composites. *Adv. Eng. Mater.* **2009**, 1039–1043. [CrossRef]

22. Fu, Z.; Schlier, L.; Travitzky, N.; Greil, P. Three-dimensional printing of SiSiC lattice truss structures. *Mater. Sci. Eng. A* **2013**, *560*, 851–856. [CrossRef]
23. Moon, J.; Caballero, A.C.; Hozer, L.; Chiang, Y.-M.; Cima, M.J. Fabrication of functionally graded reaction infiltrated SiC–Si composite by three-dimensional printing (3DP^{TM}) process. *Mater. Sci. Eng. A* **2001**, *298*, 110–119. [CrossRef]
24. Terrani, K.; Jolly, B.; Trammell, M. 3D printing of high-purity silicon carbide. *J. Am. Ceram. Soc.* **2020**, *103*, 1575–1581. [CrossRef]
25. Fleisher, A.; Zolotaryov, D.; Kovalevsky, A.; Muller-Kamskii, G.; Eshed, E.; Kazakin, M.; Popov, V.V. Reaction bonding of silicon carbides by Binder Jet 3D-Printing, phenolic resin binder impregnation and capillary liquid silicon infiltration. *Ceram. Int.* **2019**, *45*, 18023–18029. [CrossRef]
26. Halbig, M.C.; Grady, J.E.; Ramsey, J.; Patterson, C.; Santelle, T. A Fully Nonmetallic Gas Turbine Engine Enabled by Additive Manufacturing Part III: Additive Manufacturing and Characterization of Ceramic Composites. 2015. Available online: https://ntrs.nasa.gov/archive/nasa/casi.ntrs.nasa.gov/20150023455.pdf (accessed on 26 March 2020).
27. Bai, Y.; Wagner, G.; Williams, C.B. Effect of Particle Size Distribution on Powder Packing and Sintering in Binder Jetting Additive Manufacturing of Metals. *J. Manuf. Sci. Eng.* **2017**, *139*, 081019. [CrossRef]
28. Du, W.; Ren, X.; Ma, C.; Pei, Z. Binder jetting additive manufacturing of ceramics: A literature review. In Volume 14: Emerging Technologies, Materials: Genetics to Structures, Safety Engineering and Risk Analysis, Proceedings of the ASME 2017 International Mechanical Engineering Congress and Exposition, Tampa, FL, USA, 3–9 November 2017; American Society of Mechanical Engineers: New York, NY, USA, 2017.
29. Iveković, A.; Dražić, G.; Novak, S. Densification of a SiC-matrix by electrophoretic deposition and polymer infiltration and pyrolysis process. *J. Eur. Ceram. Soc.* **2011**, *31*, 833–840. [CrossRef]
30. Evans, A.G.; Wilshaw, T.R. Quasi-static solid particle damage in brittle solids—I. Observations analysis and implications. *Acta Metall.* **1976**, *24*, 939–956. [CrossRef]
31. Niihara, K. A fracture mechanics analysis of indentation-induced Palmqvist crack in ceramics. *J. Mater. Sci. Lett.* **1983**, *2*, 221–223. [CrossRef]
32. Haase, V.; Kirschstein, G.; List, H.; Ruprecht, S.; Sangster, R.; Schröder, F.; Töpper, W.; Vanecek, H.; Heit, W.; Schlichting, J.; et al. The Si-C Phase Diagram. In *Si Silicon*; Springer: Berlin/Heidelberg, Germany, 1985; pp. 1–5.
33. Yakimova, R.; Syväjärvi, M. Liquid Phase Epitaxy of Silicon Carbide. In *Liquid Phase Epitaxy of Electronic, Optical and Optoelectronic Materials*; John Wiley & Sons, Ltd.: Chichester, UK, 2007; pp. 179–201.
34. Károly, Z.; Mohai, I.; Klébert, S.; Keszler, A.; Sajó, I.E.; Szépvölgyi, J. Synthesis of SiC powder by RF plasma technique. *Powder Technol.* **2011**, *214*, 300–305. [CrossRef]
35. Li, H.; Zhang, L.; Cheng, L.; Wang, Y.; Yu, Z.; Huang, M.; Tu, H.; Xia, H. Effect of the polycarbosilane structure on its final ceramic yield. *J. Eur. Ceram. Soc.* **2008**, *28*, 887–891. [CrossRef]
36. Luo, Z.; Zhou, X.; Yu, J.; Wang, F. High-performance 3D SiC/PyC/SiC composites fabricated by an optimized PIP process with a new precursor and a thermal molding method. *Ceram. Int.* **2014**, *40*, 6525–6532. [CrossRef]
37. Hayun, S.; Paris, V.; Mitrani, R.; Kalabukhov, S.; Dariel, M.P.; Zaretsky, E.; Frage, N. Microstructure and mechanical properties of silicon carbide processed by Spark Plasma Sintering (SPS). *Ceram. Int.* **2012**, *38*, 6335–6340. [CrossRef]
38. Munro, R.G. Material Properties of a Sintered α-SiC. *J. Phys. Chem. Ref. Data* **1997**, *26*, 1195–1203. [CrossRef]
39. Lee, J.-S.; Imai, M.; Yano, T. Fabrication and mechanical properties of oriented SiC short-fiber-reinforced SiC composite by tape casting. *Mater. Sci. Eng. A* **2003**, *339*, 90–95. [CrossRef]
40. Shin, D.-W.; Park, S.S.; Choa, Y.-H.; Niihara, K. Silicon/Silicon Carbide Composites Fabricated by Infiltration of a Silicon Melt into Charcoal. *J. Am. Ceram. Soc.* **2004**, *82*, 3251–3253. [CrossRef]
41. Liu, H.; Cheng, H.; Wang, J.; Tang, G. Dielectric properties of the SiC fiber-reinforced SiC matrix composites with the CVD SiC interphases. *J. Alloy. Compd.* **2010**, *491*, 248–251. [CrossRef]
42. Liu, H.; Cheng, H.; Wang, J.; Tang, G. Effects of the single layer CVD SiC interphases on the mechanical properties of the SiCf/SiC composites fabricated by PIP process. *Ceram. Int.* **2010**, *36*, 2033–2037. [CrossRef]

© 2020 by the authors. Licensee MDPI, Basel, Switzerland. This article is an open access article distributed under the terms and conditions of the Creative Commons Attribution (CC BY) license (http://creativecommons.org/licenses/by/4.0/).

Erratum

Erratum: Polozov, I., *et al.* Fabrication of Silicon Carbide Fiber-Reinforced Silicon Carbide Matrix Composites Using Binder Jetting Additive Manufacturing from Irregularly-Shaped and Spherical Powders. *Materials* 2020, *13*, 1766

Igor Polozov [1,*], Nikolay Razumov [1], Dmitriy Masaylo [1], Alexey Silin [1], Yuliya Lebedeva [2] and Anatoly Popovich [1]

[1] Peter the Great St. Petersburg Polytechnic University, Polytechnicheskaya, 29, St. Petersburg 195251, Russia; n.razumov@onti.spbstu.ru (N.R.); dmasaylo@gmail.com (D.M.); silin8888@mail.ru (A.S.); director@immet.spbstu.ru (A.P.)
[2] Federal State Unitary Enterprise "All-Russian Scientific Research Institute of Aviation Materials" State Research Center of the Russian Federation, 17 Radio str., Moscow 105005, Russia; yulia.ananieva@gmail.com
* Correspondence: polozov_ia@spbstu.ru

Received: 19 May 2020; Accepted: 5 June 2020; Published: 9 June 2020

The authors wish to make the following correction to this paper [1]:

Change in Funding.

The authors wish to change the funding part from:

Funding: This research was supported by Russian Science Foundation grant (project No 19-79-30002).

to

Funding: This research received no external funding.

The authors would like to apologize for any inconvenience caused to the readers by these changes.

References

1. Polozov, I.; Razumov, N.; Masaylo, D.; Silin, A.; Lebedeva, Y.; Popovich, A. Fabrication of Silicon Carbide Fiber-Reinforced Silicon Carbide Matrix Composites Using Binder Jetting Additive Manufacturing from Irregularly-Shaped and Spherical Powders. *Materials* **2020**, *13*, 1766. [CrossRef]

 © 2020 by the authors. Licensee MDPI, Basel, Switzerland. This article is an open access article distributed under the terms and conditions of the Creative Commons Attribution (CC BY) license (http://creativecommons.org/licenses/by/4.0/).

Article

Additive Manufacturing of Ti-48Al-2Cr-2Nb Alloy Using Gas Atomized and Mechanically Alloyed Plasma Spheroidized Powders

Igor Polozov [1],*, Artem Kantyukov [1], Ivan Goncharov [1], Nikolay Razumov [1], Alexey Silin [1], Vera Popovich [2], Jia-Ning Zhu [2] and Anatoly Popovich [1]

[1] Peter the Great St. Petersburg Polytechnic University, Polytechnicheskaya 29, 195251 St. Petersburg, Russia; kantyukov.artem@mail.ru (A.K.); goncharov_is@spbstu.ru (I.G.); n.razumov@onti.spbstu.ru (N.R.); silin_ao@spbstu.ru (A.S.); director@immet.spbstu.ru (A.P.)

[2] Department of Materials Science and Engineering, Delft University of Technology, 2628 Delft, The Netherlands; V.Popovich@tudelft.nl (V.P.); J.Zhu-2@tudelft.nl (J.-N.Z.)

* Correspondence: polozov_ia@spbstu.ru

Received: 20 August 2020; Accepted: 4 September 2020; Published: 7 September 2020

Abstract: In this paper, laser powder-bed fusion (L-PBF) additive manufacturing (AM) with a high-temperature inductive platform preheating was used to fabricate intermetallic TiAl-alloy samples. The gas atomized (GA) and mechanically alloyed plasma spheroidized (MAPS) powders of the Ti-48Al-2Cr-2Nb (at. %) alloy were used as the feedstock material. The effects of L-PBF process parameters—platform preheating temperature—on the relative density, microstructure, phase composition, and mechanical properties of printed material were evaluated. Crack-free intermetallic samples with a high relative density of 99.9% were fabricated using 900 °C preheating temperature. Scanning electron microscopy and X-Ray diffraction analyses revealed a very fine microstructure consisting of lamellar α_2/γ colonies, equiaxed γ grains, and retained β phase. Compressive tests showed superior properties of AM material as compared to the conventional TiAl-alloy. However, increased oxygen content was detected in MAPS powder compared to GA powder (~1.1 wt. % and ~0.1 wt. %, respectively), which resulted in lower compressive strength and strain, but higher microhardness compared to the samples produced from GA powder.

Keywords: selective laser melting; additive manufacturing; titanium alloy; microstructure; mechanical alloying

---

1. Introduction

Gamma TiAl-based intermetallic alloys are attractive materials for structural high-temperature applications due to their high specific strength at room and elevated temperatures, good creep, and oxidation resistance [1]. These properties make them promising candidates for replacing nickel-based superalloys in gas turbine engines [2]. One of the most widely known γ-TiAl alloys is the Ti-48Al-2Cr-2Nb (at. %) alloy, which is currently used by General Electric for low-pressure turbine blades [3,4]. While TiAl alloys possess high strength, their poor ductility and brittleness at room temperatures severely complicate their processability by conventional manufacturing techniques and limit their application [5]. Casting followed by hot isostatic pressing (HIP) has been used to conventionally fabricate TiAl alloy parts [6]; however, this approach has its limitations in terms of high cost and design flexibility.

In recent years, additive manufacturing (AM) has been applied to manufacture various titanium alloys [7–9] as well as titanium aluminide alloys [10–12]. AM is a promising way to manufacture intermetallic alloy parts since it offers significant advantages in terms of design freedom and cost reduction compared to conventional methods. However, high cooling rates typical for laser-based powder bed fusion and directed energy deposition AM techniques lead to high residual stresses, which makes it difficult to produce crack-free intermetallic parts. Shi X. et al. [13] showed that it is not possible to produce crack-free Ti-47Al-2Cr-2Nb alloy samples by the laser powder bed fusion (L-PBF) using 200 °C preheating. The L-PBF of Ti-48Al-2Cr-2Nb alloy with 450 °C preheating resulted in severe cracking formation, indicating that higher preheating temperatures must be utilized [14]. Selective electron beam melting (SEBM) has been proved feasible in fabrication of TiAl alloys [15–17]. Utilizing an electron beam to preheat the powder bed to temperatures around 1000 °C allows one to drastically reduce residual stresses and suppress crack formation during the fabrication of TiAl alloys. One disadvantage of SEBM process is its much lower geometrical precision compared to L-PBF process. High-temperature powder bed preheating is required to produce crack-free TiAl alloys using the L-PBF process. Platform preheating temperatures of 800–1000 °C with inductive heating have been successfully used to obtain crack-free samples during the L-PBF of titanium aluminides [18,19]. However, microstructure and mechanical properties are still underexplored and require extensive characterized to evaluate the feasibility of L-PBF for manufacturing of TiAl alloys.

Another concern for further development of AM TiAl is the commercial availability of intermetallic alloys powders with properties suitable for their application in AM. Gas and plasma atomization are currently the most common processes used for the fabrication of spherical powders for L-PBF or SEBM technologies [20,21]. While atomization techniques allow manufacturing of spherical powders, these technologies are rather costly, especially considering the complex compositions of intermetallic alloys. Mechanical alloying (MA), followed by the plasma spheroidization (PS) process, is an alternative approach to obtain spherical powders with reduced costs for application in AM [22,23]. Irregular MA powders are treated in a high-temperature plasma jet, resulting in rapid melting and solidification and leading to spherical powders [24]. Application of MA plasma spheroidized (MAPS) powders in L-PBF or SEBM processes can lead to lower cost of these techniques. However, the microstructure and properties of alloys fabricated by AM from MAPS powders are yet to be evaluated and compared to gas atomized (GA) powders to assess the feasibility of such an approach.

The objective of this paper is to evaluate the feasibility of L-PBF process to fabricate crack-free Ti-48Al-2Cr-2Nb alloy from different feedstock powders and using an inductive high-temperature platform preheating. Two types of powders are used: a commercially available GA powder and an in-house produced MAPS TiAl-alloy powder. The effects of L-PBF process parameters and preheating temperature on fabricated TiAl-alloy microstructure and mechanical properties are investigated.

2. Materials and Methods

2.1. Materials

Two types of powders were used in the study to fabricate the samples by L-PBF process. The GA powder of Ti-48Al-2Cr-2Nb alloy supplied by AMC Powders (Beijing, China) and produced by electrode induction gas atomization (EIGA) had the following particle size distribution: d_{10} = 17.4 μm, d_{50} = 33.8 μm, d_{90} = 60.5 μm. Further description of the powder's composition and morphology is presented in Section 3.1. The second powder of TiAl-alloy was produced in-house by MA and PS processes. Preliminary experiments of the MA process were carried out using a Fritsch Pulverisette 4 planetary mill by milling the elemental powders of Ti, Al, Cr, and Nb (with 99.9% purity) blended with the proportion of Ti-48Al-2Cr-2Nb (at. %) alloy. A more detailed description of the MA process and characterization of the powders can be found in [25]. A dry grinding SD5 laboratory attritor produced by Union Process (Akron, OH, USA) was used for MA. The elemental powder blend was milled for 12 h in an argon atmosphere with the rotation speed of 270 rpm and 20:1 ball to powder mass ratio using stainless steel balls with 10 mm diameter. After the MA process, the powder was sieved to 0–71 μm fraction and subjected to PS using the TEK-15 system (Tekna, Sherbrooke, QC, Canada). The Ar-He gas was used as the plasma forming gas. The powder feeding rate was set to 15 g/min and the plasma torch power was 15 kW. The final particle size distribution of the MAPS powder was the following: d_{10} = 9.7 μm, d_{50} = 33.3 μm, d_{90} = 67.7 μm.

2.2. Laser Powder-Bed Fusion

The L-PBF process was carried out using AconityMIDI (Aconity3D GmbH, Herzogenrath, Germany) system. The system is equipped with a 1070 nm wavelength fiber laser with a maximum power of 1000 W. Cylindrical samples with a 10 mm diameter and 10 mm height were fabricated for further investigation. The samples were fabricated on a Ti-6Al-4V substrate, which was put on a molybdenum platform. The molybdenum substrate was inductively preheated to a set temperature, which was continuously controlled by a thermocouple under the molybdenum platform. The titanium substrate was then conductively heated by the molybdenum substrate before starting the L-PBF process. The process chamber was continuously flooded with high purity argon gas to achieve oxygen content in the chamber below 20 ppm. After the build process was finished, the platform and the samples were cooled to room temperature with a cooling rate of approximately 5 °C/min.

The platform preheating temperature was varied from 600 to 900 °C and the scanning speed (S) was varied from 650 to 1250 mm/s, while the laser power (P), hatching distance (HD), and layer thickness (L) were set to a fixed value for most of the samples. The laser spot diameter was set to ~80 µm for most of the samples. A chessboard scanning pattern with 5×5 mm^2 squares and a rotation angle of 67 °C was used for the L-PBF process. The L-PBF processing parameters used to produce the samples from the MAPS and the GA powders are shown in Table 1. The values were chosen based on the preliminary results and the published data on L-PBF of TNBV4 alloy [26].

Table 1. The L-PBF process parameters used for to produce the samples.

Parameter Set	P, W	S, mm/s	HD, mm	L, mm	VED, J/mm^3	Laser Spot Diameter, µm	Preheating Temperature, °C	Feedstock Powder
1_600	150	800	0.11	0.03	57			
2_600	150	650	0.11	0.03	70			
3_600	150	1000	0.11	0.03	45		600	
4_600	150	1250	0.11	0.03	36			
5_600	150	950	0.11	0.03	48			
6_600	150	750	0.11	0.03	61			
1_800	150	800	0.11	0.03	57			
2_800	150	650	0.11	0.03	70			
3_800	150	1000	0.11	0.03	45	80	800	GA, MAPS
4_800	150	1250	0.11	0.03	36			
5_800	150	950	0.11	0.03	48			
6_800	150	750	0.11	0.03	61			
1_900	150	800	0.11	0.03	57			
2_900	150	650	0.11	0.03	70			
3_900	150	1000	0.11	0.03	45		900	
4_900	150	1250	0.11	0.03	36			
5_900	150	950	0.11	0.03	48			
6_900	150	750	0.11	0.03	61			
Large spot	850	330	0.45	0.09	64	500	800	GA

One sample was fabricated using GA powder and a large laser spot diameter of 500 µm (further denoted as "Large spot") and increased laser power and layer thickness values, while the volume energy density (VED) was at a similar value as for the samples produced with 80 µm spot diameter (see Table 1). The larger spot was used in order to investigate how the microstructure of a TiAl-alloy can be varied by using an increased laser spot diameter coupled with high laser power and increased layer thickness. As demonstrated in the previous study, L-PBF processing with increased power energy can result in a strongly textured coarse microstructure, as shown in the case of Inconel 718 alloy [27,28].

2.3. Characterization

The as-fabricated samples were cut and polished along the build direction (BD) for the microstructural characterization. Mira 3 LMU (TESCAN, Brno, Czech Republic) scanning electron microscope (SEM) in the backscattered electrons (BSE) mode was utilized to evaluate the microstructures of non-etched samples. Energy Dispersive Spectroscopy (EDS) was used for the chemical analysis of the samples and powders on the polished cross-sections.

The phase composition of the powders and the fabricated samples was analyzed with a Bruker D8 Advance X-ray diffraction (XRD) (Bruker, Germany) using Cu-Kα (λ = 1.5418 Å) irradiation.

The relative density was measured by a standard metallographic technique, which includes taking a minimum of five different locations of the polished samples with an optical microscope (OM) Leica DMI5000 () at 50× magnification. The OM images were then used to isolate the pores from the bulk material (Leica, Wetzlar, Germany) using ImageJ software. The calculated fraction of the image defined as bulk materials was used as the relative density value. General Electric (Boston, MA, USA) Phoenix Vtomex Computed Tomography (CT) System was used for X-ray microtomography analyses (CT) of the samples with a voxel size of 10 μm. Avizo software was used to visualize the CT-data and evaluate the porosity of the samples.

The oxygen content in the powder and the fabricated samples was measured using the inert-gas fusion-infrared (IGF) method with a LECO TC-500 analyzer (LECO, St Joseph, MI, USA).

The microhardness of the samples was measured using a Buehler VH1150 testing machine with 300 g load and 10 s dwell time.

L-PBF process parameter sets that produced the sample with the highest relative density were then used to fabricate cylindrical samples with 4 mm diameter and 20 mm height for the compression tests. The samples were cut by electrical discharge machining to achieve 5 mm height. Room temperature compression tests were performed using a universal testing machine (Zwick/Roell Z100, Ulm, Germany) with a strain rate of 0.1 mm/min. A minimum of three samples per point were tested.

3. Results and Discussion

3.1. Powder Characterization

The particle size distribution of gas atomized (GA) powder was found to be d_{10} = 17.4 μm, d_{50} = 33.8 μm, d_{90} = 60.5 μm. As can be seen in Figure 1, the GA particles have a spherical shape and dendritic surface morphology, while the cross-section shows a typical for the GA process dendritic microstructure [29,30]. The chemical composition of the powder is shown in Table 2.

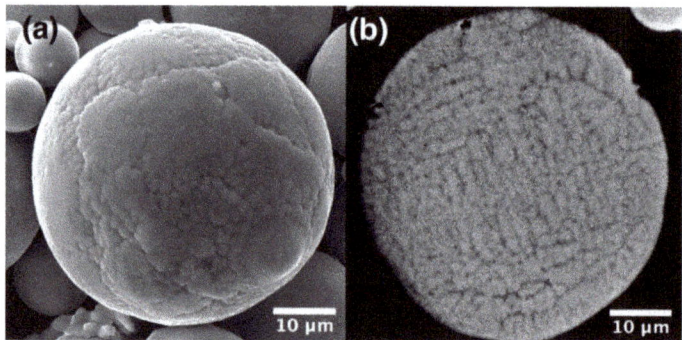

Figure 1. Scanning electron microscope (SEM) images of the gas atomized powder (GA) showing (**a**) surface morphology and (**b**) cross-section of a particle.

Table 2. The chemical composition of mechanically alloyed plasma spheroidized (MAPS) and GA powders.

Powder	Measured by EDS				Measured by IGF
	Ti, at. %	Al, at. %	Nb, at. %	Cr, at. %	O, wt. %
MAPS	51.6 ± 0.3	43.7 ± 0.3	2.0 ± 0.1	2.2 ± 0.1	1.1 ± 0.1
GA	50.2 ± 0.2	45.7 ± 0.3	2.1 ± 0.1	2.1 ± 0.1	0.07 ± 0.01

The particle size distribution of mechanical alloyed plasma spheroidized (MAPS) powder was found to be d_{10} = 9.7 μm, d_{50} = 33.3 μm, d_{90} = 67.7 μm. The particles also have a spherical shape and dendritic surface morphology (Figure 2a). A significant difference from the GA powder is the presence of small oxides inside the particles, which can be seen as black precipitates in the cross-section image (Figure 2b). The presence of oxides can be attributed to an increased oxygen content as shown in Table 2. The oxygen pickup most likely occurred during the MA process. Al was partially lost during the PS process. Thus, further optimization of the process may be carried out to obtain a proper chemical composition.

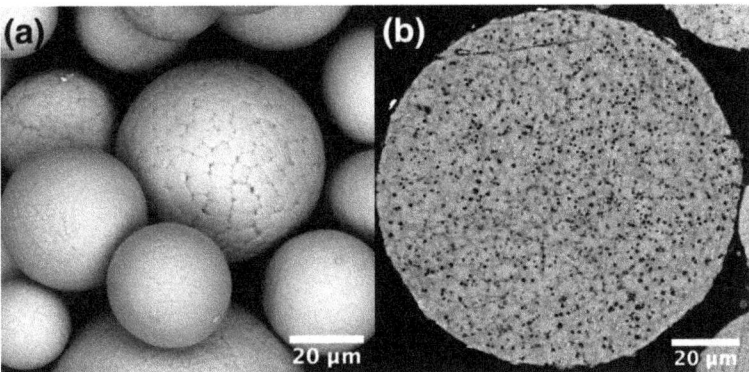

Figure 2. SEM images of the mechanically alloyed plasma spheroidized (MAPS) powder showing (**a**) surface morphology and (**b**) cross-section of a particle.

The XRD pattern of the GA powder shows peaks corresponding to α/α_2-phase with a hexagonal lattice and a small peak corresponding to β-phase with a body-centered cubic (BCC) lattice (Figure 3). The GA process is characterized by high cooling rates during the solidification of powders preserving the α/α_2-phase. The phase composition of MAPS powder is characterized by the presence of α_2 (Ti3Al-based) and γ (TiAl-based) phases. The presence of the ordered α_2-phase instead of α-phase can be confirmed by identifying the most intense superlattice α_2 peak [31], which is seen in the case of the MAPS powder. The difference in the phase composition between GA and MAPS powders might be the result of Al and O content differences. As shown in [32], an increase of oxygen content can decrease the volume fraction of γ-phase and increase the volume fraction of hexagonal α/α_2-phase.

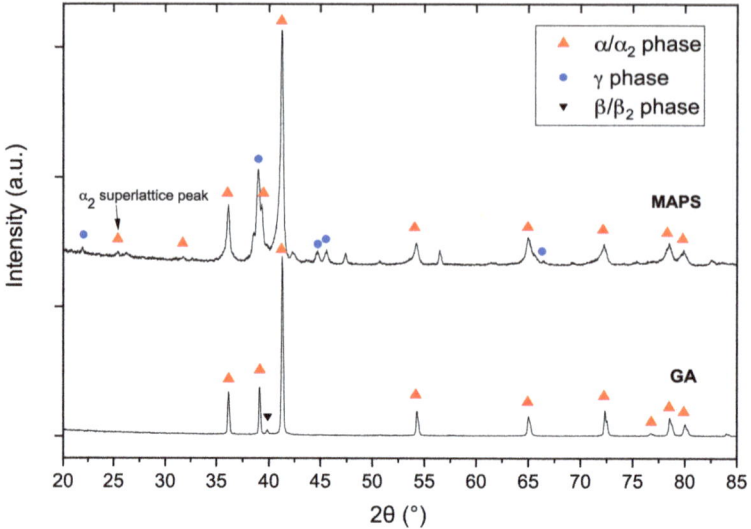

Figure 3. X-ray diffraction (XRD) patterns of MAPS and GA powders.

3.2. Densification via Laser Powder-Bed Fusion

Figure 4 shows the effect of platform preheating temperature on the formation of cracks in samples fabricated from the MAPS powder. As can be seen in Figure 4a, severe cracking is observed when a relatively low preheating temperature of 600 °C was used during the L-PBF process. Increasing the preheating temperature to 800 °C significantly reduced the number of cracks; however, occasional horizontal cracks perpendicular to the BD were still present. Further increase in the preheating temperature up to 900 °C resulted in elimination of cracks. This is in agreement with the brittle-ductile transition temperature (BDTT) of the TiAl-alloy, which is around 750–780 °C [33]. Increased ductility above the BDDT allows the material to accommodate high stresses during L-PBF and avoid cracking. Similar results were obtained in [18] when the Ti-44.8Al–6Nb–1.0Mo–0.1B (at. %) powder was processed by L-PBF at 800 °C platform preheating. It should be noted that occasional horizontal cracks were still present on the edges of the specimen, likely due to a high thermal gradient (Figure 4b). Similar results in terms of cracks formation were obtained for the GA powder. Further investigation of the relative density was carried out only on the samples fabricated at 800 and 900 °C preheating temperatures.

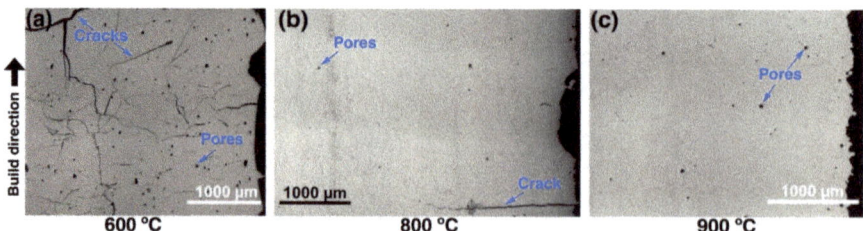

Figure 4. Optical images of the cross-sections of the samples fabricated from MAPS powder. The images show representative cross-sections of the samples obtained using different preheating temperatures: (**a**) 600 °C (sample 1_600), (**b**) 800 °C (sample 1–800), (**c**) 900 °C (sample 1_900).

Figures 5 and 6 show the effect of scanning speed on the relative density of the samples fabricated using the MAPS and the GA powders, respectively. In the case of the MAPS powder, the highest relative density (99.75 ± 0.05%) was achieved at 650 mm/s scanning speed (which corresponds to 70 J/mm^3 VED) for both 800 and 900 °C preheating temperatures. Using the GA powder resulted in an overall higher relative density values (all values were higher than 99.6%). The highest relative density of 99.94 ± 0.05% was obtained at 800 °C preheating temperature and 950 mm/s scanning speed which corresponds to 48 J/mm³ VED. In general, a higher preheating temperature resulted in a slightly higher porosity, which suggests that some overheating might have taken place during the L-PBF and led to formation of keyhole pores. At the same time, applying high scanning speed of 1250 mm/s resulted in an increased porosity for both powders, which might be the result of insufficient melting or melt pool instability [34,35].

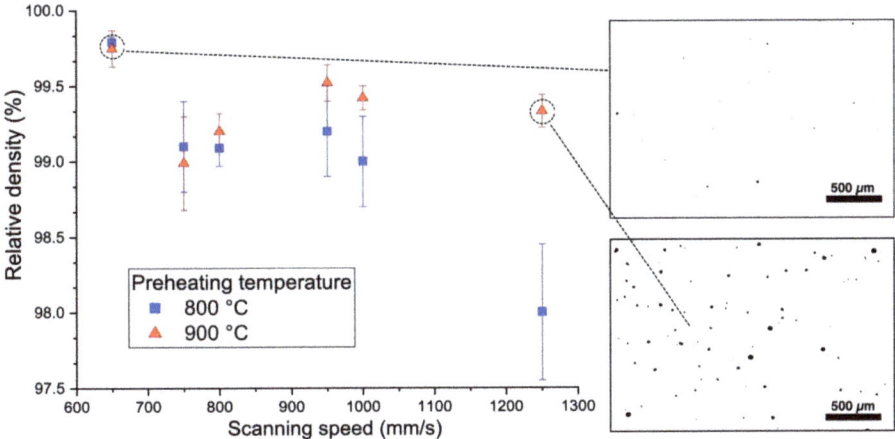

Figure 5. Effect of scanning speed and platform preheating temperature on the relative density of the samples fabricated from MAPS powder with cross-section images showing typical pore distribution.

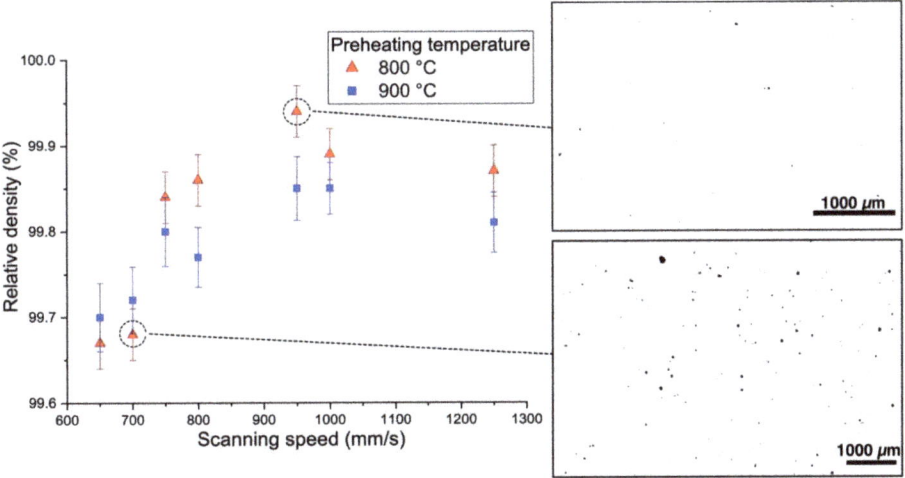

Figure 6. Effect of scanning speed and platform preheating temperature on the relative density of the samples fabricated from GA powder with cross-section images showing typical pore distribution.

The samples fabricated from the MAPS and the GA powders having the highest relative densities were taken further for the CT-investigation. The results (Figure 7) showed that the median size of the pores was around 40 μm for the MAPS sample and around 15 μm for the GA sample. The porosity volume was visibly lower in the case of the GA sample compared to the MAPS sample, which is compliant with the metallographic measurements. The pores were found to be mostly spherical, suggesting that these are gas pores formed due to entrapped gas originating from the argon gas or melt pool vaporization [36]. A few lack-of-fusion defects were also found for both samples, indicating poor bonding defects or incomplete melting of some powders [37]. According to the CT-results, the GA sample had porosity volume less than 0.1%, while the MAPS sample had porosity volume of around 0.3%. Higher porosity in MAPS samples might be attributed to an increased oxygen in the feedstock powder, as well as the initial porosity from the powder. This correlates with the results obtained by Li et al. [38], where they showed that a powder with lower oxygen content resulted in better densification during L-PBF of 316L alloy.

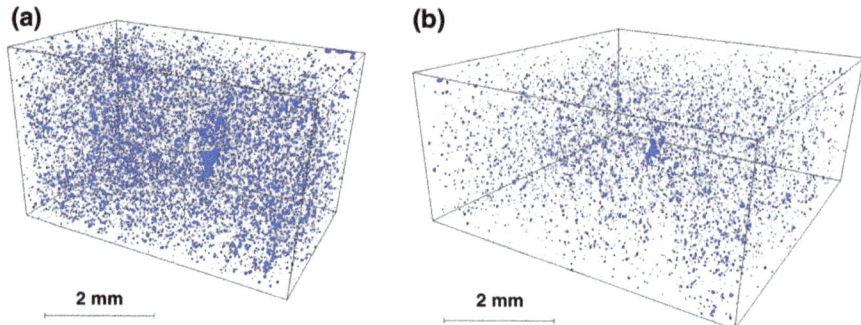

Figure 7. Computer tomographic reconstruction of the porosity volume in the samples produced from (a) MAPS powder (sample 2_800) and (b) GA powder (sample 5_800).

3.3. Microstructural Characterization

Figure 8 shows microstructure of the TiAl-alloy fabricated from MAPS powder at 800 °C (Figure 8a) and 900 °C (Figure 8b) preheating temperatures. According to the Ti-Al phase diagram [39], these temperatures correspond to the $\alpha_2+\gamma$ phase field. In both cases, the alloy has a very fine duplex microstructure consisting of lamellar α_2/γ colonies (gray), equiaxed γ grains (dark gray), and residual β phase (white) as confirmed by the XRD results (Figure 9). Retained β-phase was also found in the microstructure of the SEBM-processed Ti-48Al-2Cr-2Nb alloy in the paper [10]. While the Ti-48Al-2Cr-2Nb is an α-solidifying alloy, high cooling rates during the L-PBF process can induce solidification of a metastable β-phase. As shown in Table 3, there is Al loss of about 2–3 at. % in the samples fabricated from MAPS powder compared to the initial powder. Reducing the Al concentration to about 40–41 at. % leads to the solidification through the Ti-rich side of the peritectic reaction and formation of the β-phase. The retained β-phase in the TiAl-alloy can increase its ductility, however the strength at room and elevated temperatures can be worsened due to refractory elements segregation and reduced solid solution strengthening in the lamellar regions [40]. Thus, a subsequent heat treatment should be considered to transform the metastable β-phase in the TiAl alloy.

According to the XRD results (Figure 10), the alloy has some oxides corresponding to the TiO phase. The oxides mostly likely originated from the initial powder; however, some oxygen pickup during the L-PBF process could take place as well. The oxygen measured by LECO analysis showed that samples fabricated from MAPS powder had oxygen content around 1.3 wt. %, while samples fabricated from GA powder had 10 times lower oxygen content (~0.14 wt. %).

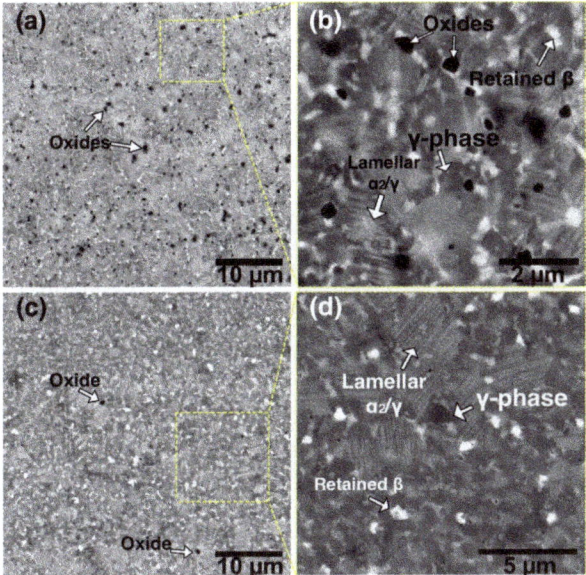

Figure 8. BSE-SEM images showing microstructures of the samples fabricated from the MAPS powder at (**a**,**b**) 800 °C (sample 2_800) and (**c**,**d**) 900 °C preheating temperature (sample 2_900).

Figure 9 shows microstructures of the samples fabricated from GA powder at 800 °C and 900 °C preheating temperatures. The obtained microstructure is similar to MAPS samples and mainly consists of lamellar α_2/γ colonies, equiaxed γ grains, and the retained β phase. Crescent-shaped melt pool boundaries with a width of about 80–90 µm can be found in the microstructure as shown in Figure 9a. There were no oxide precipitates found in the case of GA powder due to relatively low oxygen content in the initial powder. Compared to the MAPS samples, the amount of retained β phase is visibly lower in GA samples since their Al content is higher (as can be seen from Tables 3 and 4) and the solidification is expected to take place mainly through the α phase region. However, when the preheating temperature was increased to 900 °C from 800 °C the amount of β phase visibly increased.

The microstructure of the samples fabricated from GA powder shows fewer lamellar α_2/γ colonies compared to the MAPS samples. This could due to lower oxygen content in the GA samples since a decreased oxygen content leads to an increased γ-phase volume fraction in TiAl-alloys [32]. XRD results (Figure 10) confirmed that samples fabricated from GA powder consist mainly of γ phase (TiAl) and α_2 (Ti$_3$Al) phase.

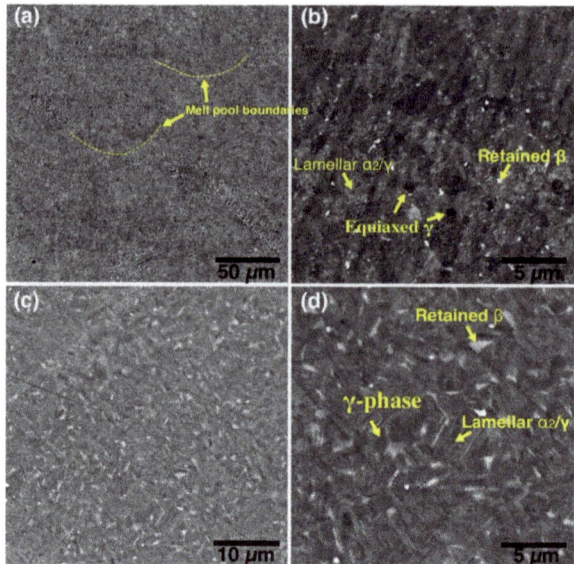

Figure 9. Backscattered electrons (BSE) SEM images showing microstructures of the samples fabricated from GA powder at (**a**,**b**) 800 °C (sample 5_800) and (**c**,**d**) 900 °C preheating temperature (sample 5_900).

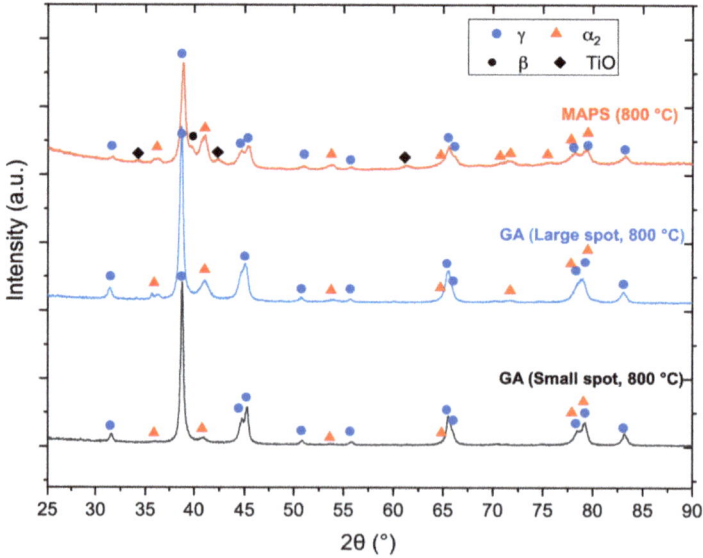

Figure 10. XRD patterns of the samples fabricated from GA and MAPS powders.

Figure 11 shows the microstructure of the "Large spot" sample fabricated from the GA powder using a laser spot diameter of about 500 μm. In this case, the melt pool width was roughly about 500–600 μm as can be seen in Figure 11a. Larger melt pool during the L-PBF process is expected to result in lower cooling rates and, subsequently, a coarser microstructure [41]. The obtained microstructure is highly inhomogeneous and consists of various layered regions. There are coarse columnar γ grains inside the melt pool growing towards the middle of the melt pool, as can be seen in Figure 11b. A fine-grained zone can be seen underneath the melt pool boundary, which likely formed during recrystallization in a heat-affected zone (HAZ). The L-PBF involves repetitive heating and cooling resulting in a complex thermal history. This leads to intrinsic non-uniform heat treatment during the L-PBF process. A similar alternative-band microstructure of the Ti-47Al-2Cr-2Nb alloy was observed after direct laser deposition in [42] and the authors attributed it to the effect of cyclic heat treatment derived from multiple laser exposure. The deposition and laser exposure of the next layers leads to heating the underlying solidified material. In this HAZ the material reaches α-transus temperature, then cools down into α+γ region resulting in a refined microstructure [10].

Al-rich and Al-poor regions can be distinguished from the BSE-SEM images of the "Large spot" sample. Al-rich zone is seen at the bottom of a melt pool as a dark band while the Al depletion is observed at the top of the melt pool, as can be seen in Figure 11d. It suggests that the Al loss occurred mostly closer to the surface of a melt pool. The overall Al loss in "Large spot" sample was around 3 at. % compared to the initial powder (Table 4). Al segregation in the L-PBF-processed Ti-6Al-4V alloy samples was observed in [43] resulting in dark bands at the bottom of a melt pool. Similarly, SEBM resulted in a banded, inhomogeneous microstructure of a γ-TiAl based alloy due to the vaporization of Al [44]. Al-rich regions consist of lamellar α_2/γ colonies as can be seen in Figure 11c, while Al-poor regions have lamellar colonies surrounded by the retained β phase as shown in Figure 11e. The differences in microstructures between Al-rich and Al-poor areas are in a good agreement with the Ti-Al phase diagram [45].

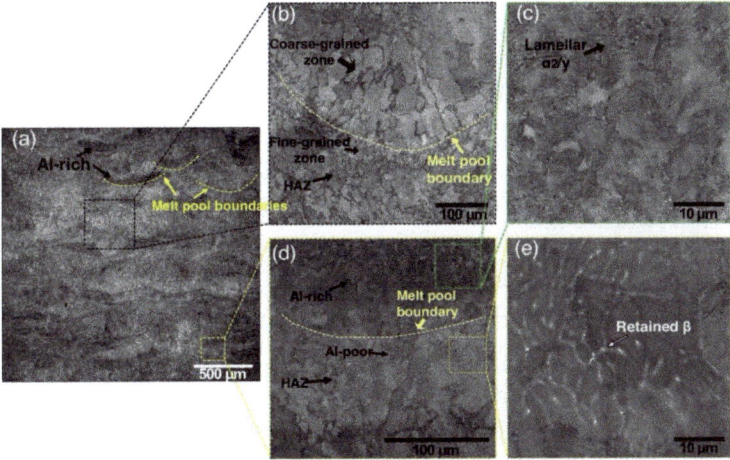

Figure 11. BSE-SEM images showing the microstructure variation in the "Large spot" sample produced from GA powder: (**a**) a general view, (**b**) a magnified area showing a melt pool section, (**c**) Al-rich area, (**d**) a melt pool boundary, (**e**) Al-poor area

Table 3. The chemical composition (in at. %) of the samples fabricated from MAPS powder at different preheating temperatures.

Preheating Temperature, °C	Ti	Al	Nb	Cr	O
600	55.2 ± 0.1	40.4 ± 0.1	2.0 ± 0.1	2.2 ± 0.1	–
800	53.8 ± 0.2	41.6 ± 0.2	2.3 ± 0.1	2.0 ± 0.2	0.14 ± 0.02
900	54.3 ± 0.1	40.9 ± 0.1	2.1 ± 0.2	2.2 ± 0.1	0.13 ± 0.02

Table 4. The chemical composition (in at. %) of the samples fabricated from GA powder using different laser spot diameters.

Sample	Ti	Al	Nb	Cr
Small spot, 80 µm (Sample 5_800)	50.5 ± 0.1	45.3 ± 0.2	2.1 ± 0.1	2.1 ± 0.1
Large spot, 500 µm	52.9 ± 0.2	42.3 ± 0.3	2.0 ± 0.1	2.2 ± 0.1

3.4. Mechanical Properties

Table 5 presents the results of room temperature compressive tests of the samples fabricated using GA and MAPS powders. High ultimate compressive strength values were obtained for both GA and MAPS samples, which were 2277 ± 71 MPa and 1910 ± 37 MPa, respectively. The compressive strain values were around 32%–35% for the GA samples and 15%–17% for the MAPS. Using the GA powder resulted in both higher compressive strength and strain. This can be explained by an increased oxygen content in MAPS powder, which is known to embrittle TiAl-alloys [46], as well as a higher β-phase content in MAPS samples, which is softer than intermetallic γ and α_2 phases [47] and can reduce the strength of the alloy. The overall room temperature compressive performance of the samples fabricated by L-PBF with a high-temperature preheating showed promising results and exceeded the compressive strength of SEBM samples and showed significantly better results than the samples fabricated by L-PBF with a non-preheated platform [48]. The GA samples also showed superior compressive performance compared to the conventional TiAl-alloy [49] and SEBM Ti-48Al-2Cr-2Nb alloy [50].

Table 5. Comparison of room temperature mechanical properties of TiAl alloy manufactured by different processes.

Material	Condition	Ultimate Compressive Strength, MPa	Compressive Strain, %
Ti-48Al-2Cr-2Nb (GA powder, this study)	As-fabricated, 900 °C preheating	2277 ± 71	32–35
Ti-48Al-2Cr-2Nb (MAPS powder, this study)	As-fabricated, 900 °C preheating	1910 ± 37 MPa	15–17
Ti-48Al-2Cr-2Nb (SEBM) [48]	As-fabricated	1800	40
Ti-48Al-2Cr-2Nb (L-PBF) [48]	As-fabricated, no preheating	612 ± 56	1,98 ± 0.55
Ti-48Al-2Cr-2Nb (SEBM) [50]	Heat-treated	2068	25
Ti-48Al-2Cr-2Nb (casted) [50]	As-fabricated	1153	~6

The microhardness variation over a range of 1.5 mm along the BD was measured for different samples, as shown in Figure 12. All samples demonstrated a fluctuating microhardness profile, which is associated with the microstructural inhomogeneity. A similar microhardness variation along the BD was obtained in [42] for the direct laser deposited Ti-47Al-2Cr-2Nb alloy, with an average value of 400 HV. The highest mean microhardness values were obtained for the samples fabricated from MAPS powder: 584 HV and 552 HV for 800 °C and 900 °C preheating temperatures, respectively. This correlates well with the oxygen content in the samples: higher microhardness was obtained for the samples with higher oxygen content. As shown in [51] for the Ti-48Al-2Cr-2Nb alloy, an increase in oxygen content increases α_2 phase volume fraction that has higher hardness compared to γ phase. At the same time, decreased Al content in MAPS samples could prevent the formation of γ-phase, resulting in a higher α_2-phase volume fraction and higher microhardness.

The samples fabricated from GA powder showed mean microhardness around 440 HV, which is higher than as-fabricated SEBM (253 HV) and conventional (371 HV) TiAl-alloy [50]. Higher microhardness might be attributed to a very fine microstructure in the case of the L-PBF. High microhardness can be beneficial for wear performance [52]; however, it also suggests increased brittleness and reduced ductility [51]. Thus, a more complex mechanical properties characterization is recommended.

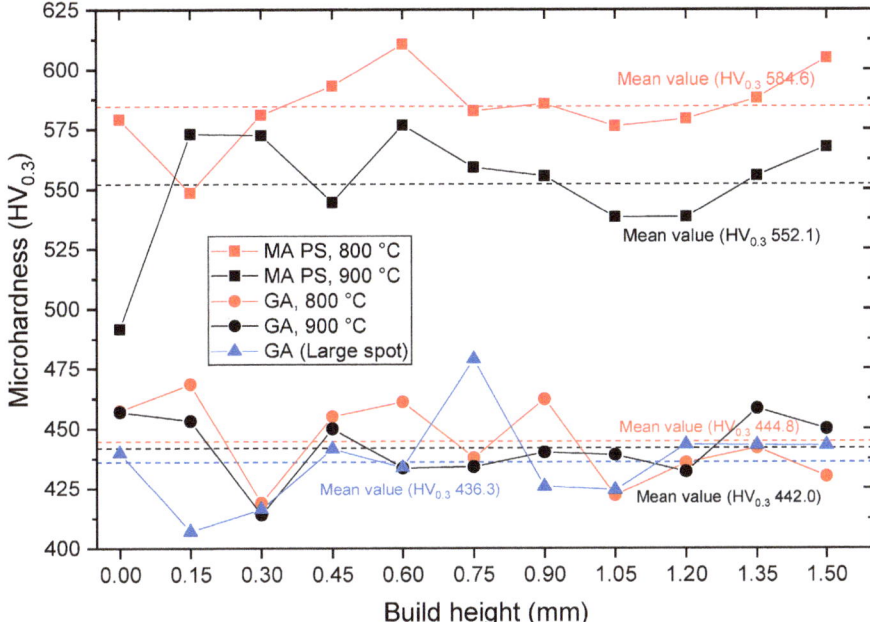

Figure 12. Microhardness variation along the build direction of the samples manufactured from MAPS powder and GA powder. The dashed lines correspond to the mean values: red—for the samples fabricated at 800 °C, black—at 900 °C, blue—using the large laser spot size.

4. Conclusions

In this work it was demonstrated that pre-alloyed spherical GA and MAPS powders of TiAl-based alloy can be used to fabricate crack-free samples using the L-PBF process with high-temperature platform preheating. The following main conclusions were drawn:

1. Crack-free samples were fabricated with 900 °C platform preheating temperature. The highest relative density of 99.9% was obtained with GA powder at 48 J/mm^3 VED and 99.7% with MAPS powder at 70 J/mm^3.

2. Very fine microstructures consisting of lamellar α_2/γ colonies, equiaxed γ grains, and retained β phase were obtained in all samples. Al loss during the L-PBF process led to the shift in the solidification route and resulted in the formation of the retained β phase. Increased oxygen content in the initial powder led to the formation of small oxides and an increased α_2 volume fraction.

3. Using an increased energy input during the L-PBF with a large laser spot size resulted in an inhomogeneous microstructure consisting of Al-rich and Al-poor regions. Regions with lamellar α_2/γ colonies and equiaxed α_2 grains surrounded with retained β phase depending on the Al concentration were found in the microstructure. There were coarse columnar and refined γ grains as a result of repetitive heating and cooling during the L-PBF.

4. The fabricated samples showed high ultimate compressive strength and strain values. The samples fabricated from GA powder demonstrated superior compressive performance compared to the samples from the MAPS powder. Both alloys showed superior compressive properties compared to the conventional TiAl-alloy.

5. The samples fabricated from the MAPS powder had higher microhardness due to the increased oxygen content and α_2 volume fraction.

6. Further investigation will be focused on a more detailed characterization of mechanical properties at room and elevated temperatures including tensile behavior, as well as the effects of heat treatment on microstructure and properties of the TiAl-alloy.

Author Contributions: Conceptualization, I.P. and N.R.; Investigation, I.P., A.K., I.G., A.S. and J.-N.Z.; Methodology, I.P., I.G. and N.R.; Project administration, A.P.; Supervision, I.P., V.P. and A.P.; Writing—original draft, I.P.; Writing—review and editing, V.P. All authors have read and agreed to the published version of the manuscript.

Funding: This research was supported by Russian Science Foundation grant (project No 19-79-30002).

Conflicts of Interest: The authors declare no conflict of interest.

References

1. Guyon, J.; Hazotte, A.; Wagner, F.; Bouzy, E. Recrystallization of coherent nanolamellar structures in Ti48Al2Cr2Nb intermetallic alloy. *Acta Mater.* **2016**, *103*, 672–680. [CrossRef]
2. Bartolotta, P.; Barrett, J.; Kelly, T.; Smashey, R. The use of cast Ti–48Al–2Cr–2Nb in jet engines. *JOM* **1997**, *49*, 48–50. [CrossRef]
3. Lin, J.; Xu, X.; Zhang, L.; Liang, Y.; Xu, Y.; Hao, G. Alloy Design Concepts for Wrought High Temperature TiAl Alloys. In *Gamma Titanium Aluminide Alloys 2014*; John Wiley & Sons, Inc.: Hoboken, NJ, USA, 2014; pp. 1–7.
4. Djanarthany, S.; Viala, J.-C.; Bouix, J. An overview of monolithic titanium aluminides based on Ti3Al and TiAl. *Mater. Chem. Phys.* **2001**, *72*, 301–319. [CrossRef]
5. Gupta, R.K.; Pant, B.; Sinha, P.P. Theory and Practice of γ + α2 Ti Aluminide: A Review. *Trans. Indian Inst. Met.* **2014**, *67*, 143–165. [CrossRef]
6. Wu, X. Review of alloy and process development of TiAl alloys. *Intermetallics* **2006**, *14*, 1114–1122. [CrossRef]
7. Huang, S.; Sing, S.L.; de Looze, G.; Wilson, R.; Yeong, W.Y. Laser powder bed fusion of titanium-tantalum alloys: Compositions and designs for biomedical applications. *J. Mech. Behav. Biomed. Mater.* **2020**, *108*, 103775. [CrossRef]
8. Fan, H.; Yang, S. Effects of direct aging on near-alpha Ti–6Al–2Sn–4Zr–2Mo (Ti-6242) titanium alloy fabricated by selective laser melting (SLM). *Mater. Sci. Eng. A* **2020**, *788*, 139533. [CrossRef]
9. Murr, L.E.; Quinones, S.A.; Gaytan, S.M.; Lopez, M.I.; Rodela, A.; Martinez, E.Y.; Hernandez, D.H.; Martinez, E.; Medina, F.; Wicker, R.B. Microstructure and mechanical behavior of Ti-6Al-4V produced by rapid-layer manufacturing, for biomedical applications. *J. Mech. Behav. Biomed. Mater.* **2009**, *2*, 20–32. [CrossRef]
10. Schwerdtfeger, J.; Körner, C. Selective electron beam melting of Ti-48Al-2Nb-2Cr: Microstructure and aluminium loss. *Intermetallics* **2014**, *49*, 29–35. [CrossRef]
11. Juechter, V.; Franke, M.M.; Merenda, T.; Stich, A.; Körner, C.; Singer, R.F. Additive manufacturing of Ti-45Al-4Nb-C by selective electron beam melting for automotive applications. *Addit. Manuf.* **2018**, *22*, 118–126. [CrossRef]
12. Polozov, I.; Sufiiarov, V.; Kantyukov, A.; Popovich, A. Selective Laser Melting of Ti2AlNb-based intermetallic alloy using elemental powders: Effect of process parameters and post-treatment on microstructure, composition, and properties. *Intermetallics* **2019**, *112*, 106554. [CrossRef]
13. Shi, X.; Wang, H.; Feng, W.; Zhang, Y.; Ma, S.; Wei, J. The crack and pore formation mechanism of Ti–47Al–2Cr–2Nb alloy fabricated by selective laser melting. *Int. J. Refract. Met. Hard Mater.* **2020**, 105247. [CrossRef]
14. Doubenskaia, M.; Domashenkov, A.; Smurov, I.; Petrovskiy, P. Study of Selective Laser Melting of intermetallic TiAl powder using integral analysis. *Int. J. Mach. Tools Manuf.* **2018**, *129*, 1–14. [CrossRef]
15. Baudana, G.; Biamino, S.; Klöden, B.; Kirchner, A.; Weißgärber, T.; Kieback, B.; Pavese, M.; Ugues, D.; Fino, P.; Badini, C. Electron Beam Melting of Ti-48Al-2Nb-0.7Cr-0.3Si: Feasibility investigation. *Intermetallics* **2016**, *73*, 43–49. [CrossRef]
16. Murr, L.E.; Gaytan, S.M.; Ceylan, A.; Martinez, E.; Martinez, J.L.; Hernandez, D.H.; Machado, B.I.; Ramirez, D.A.; Medina, F.; Collins, S. Characterization of titanium aluminide alloy components fabricated by additive manufacturing using electron beam melting. *Acta Mater.* **2010**, *58*, 1887–1894. [CrossRef]
17. Baudana, G.; Biamino, S.; Ugues, D.; Lombardi, M.; Fino, P.; Pavese, M.; Badini, C. Titanium aluminides for aerospace and automotive applications processed by Electron Beam Melting: Contribution of Politecnico di Torino. *Met. Powder Rep.* **2016**, *71*, 193–199. [CrossRef]
18. Gussone, J.; Hagedorn, Y.-C.; Gherekhloo, H.; Kasperovich, G.; Merzouk, T.; Hausmann, J. Microstructure of γ-titanium aluminide processed by selected laser melting at elevated temperatures. *Intermetallics* **2015**, *66*, 133–140. [CrossRef]

19. Caprio, L.; Demir, A.G.; Chiari, G.; Previtali, B. Defect-free laser powder bed fusion of Ti–48Al–2Cr–2Nb with a high temperature inductive preheating system. *J. Phys. Photonics* **2020**, *2*, 024001. [CrossRef]
20. Dietrich, S.; Wunderer, M.; Huissel, A.; Zaeh, M.F. A New Approach for a Flexible Powder Production for Additive Manufacturing. *Procedia Manuf.* **2016**, *6*, 88–95. [CrossRef]
21. Sun, P.; Fang, Z.Z.; Zhang, Y.; Xia, Y. Review of the Methods for Production of Spherical Ti and Ti Alloy Powder. *JOM* **2017**, *69*, 1853–1860. [CrossRef]
22. Polozov, I.; Razumov, N.; Makhmutov, T.; Silin, A.; Kim, A.; Popovich, A. Synthesis of titanium orthorhombic alloy spherical powders by mechanical alloying and plasma spheroidization processes. *Mater. Lett.* **2019**, *256*, 126615. [CrossRef]
23. Goncharov, I.S.; Razumov, N.G.; Silin, A.O.; Ozerskoi, N.E.; Shamshurin, A.I.; Kim, A.; Wang, Q.S.; Popovich, A.A. Synthesis of Nb-based powder alloy by mechanical alloying and plasma spheroidization processes for additive manufacturing. *Mater. Lett.* **2019**, *245*, 188–191. [CrossRef]
24. Ozerskoi, N.E.; Wang, Q.S. Obtaining Spherical Powders of Grade 5 Alloy for Application in Selective Laser Melting Technology. *Key Eng. Mater.* **2019**, *822*, 304–310. [CrossRef]
25. Polozov, I.; Popovich, V.; Razumov, N.; Makhmutov, T.; Popovich, A. Gamma-Titanium Intermetallic Alloy Produced by Selective Laser Melting Using Mechanically Alloyed and Plasma Spheroidized Powders. In *The Minerals, Metals & Materials Society (eds) TMS 2020 149th Annual Meeting & Exhibition Supplemental Proceedings*; The Minerals, Metals & Materials Series; Springer: Pittsburg, CA, USA, 2020; pp. 375–383.
26. Gussone, J.; Garces, G.; Haubrich, J.; Stark, A.; Hagedorn, Y.C.; Schell, N.; Requena, G. Microstructure stability of γ-TiAl produced by selective laser melting. *Scr. Mater.* **2017**, *130*, 110–113. [CrossRef]
27. Popovich, V.A.; Borisov, E.V.; Popovich, A.A.; Sufiiarov, V.S.; Masaylo, D.V.; Alzina, L. Functionally graded Inconel 718 processed by additive manufacturing: Crystallographic texture, anisotropy of microstructure and mechanical properties. *Mater. Des.* **2017**, *114*, 441–449. [CrossRef]
28. Popovich, V.A.; Borisov, E.V.; Sufiyarov, V.S.; Popovich, A.A. Tailoring the Properties in Functionally Graded Alloy Inconel 718 Using Additive Technologies. *Met. Sci. Heat Treat.* **2019**, *60*, 701–709. [CrossRef]
29. Fuchs, G.E.; Hayden, S.Z. Microstructural evaluation of as-solidified and heat-treated γ-TiAl based powders. In *High Temperature Aluminides and Intermetallics*; Elsevier: Amsterdam, The Netherlands, 1992; pp. 277–282.
30. Mullis, A.M.; Farrell, L.; Cochrane, R.F.; Adkins, N.J. Estimation of Cooling Rates During Close-Coupled Gas Atomization Using Secondary Dendrite Arm Spacing Measurement. *Metall. Mater. Trans. B* **2013**, *44*, 992–999. [CrossRef]
31. Guyon, J.; Hazotte, A.; Bouzy, E. Evolution of metastable α phase during heating of Ti48Al2Cr2Nb intermetallic alloy. *J. Alloys Compd.* **2016**, *656*, 667–675. [CrossRef]
32. Lapin, J.; Gabalcová, Z. The effect of oxygen content and cooling rate on phase transformations in directionally solidified intermetallic Ti-46Al-8Nb alloy. *Kov. Mater.* **2008**, *46*, 185–195.
33. Chen, R.; Wang, Q.; Yang, Y.; Guo, J.; Su, Y.; Ding, H.; Fu, H. Brittle–ductile transition during creep in nearly and fully lamellar high-Nb TiAl alloys. *Intermetallics* **2018**, *93*, 47–54. [CrossRef]
34. Kuo, C.N.; Chua, C.K.; Peng, P.C.; Chen, Y.W.; Sing, S.L.; Huang, S.; Su, Y.L. Microstructure evolution and mechanical property response via 3D printing parameter development of Al–Sc alloy. *Virtual Phys. Prototyp.* **2020**, *15*, 120–129. [CrossRef]
35. Gunenthiram, V.; Peyre, P.; Schneider, M.; Dal, M.; Coste, F.; Koutiri, I.; Fabbro, R. Experimental analysis of spatter generation and melt-pool behavior during the powder bed laser beam melting process. *J. Mater. Process. Technol.* **2018**, *251*, 376–386. [CrossRef]
36. Voisin, T.; Calta, N.P.; Khairallah, S.A.; Forien, J.-B.; Balogh, L.; Cunningham, R.W.; Rollett, A.D.; Wang, Y.M. Defects-dictated tensile properties of selective laser melted Ti-6Al-4V. *Mater. Des.* **2018**, *158*, 113–126. [CrossRef]
37. Zhang, B.; Li, Y.; Bai, Q. Defect Formation Mechanisms in Selective Laser Melting: A review. *Chin. J. Mech. Eng.* **2017**, *30*, 515–527. [CrossRef]
38. Li, R.; Shi, Y.; Wang, Z.; Wang, L.; Liu, J.; Jiang, W. Densification behavior of gas and water atomized 316L stainless steel powder during selective laser melting. *Appl. Surf. Sci.* **2010**, *256*, 4350–4356. [CrossRef]
39. Wei, D.-X.; Koizumi, Y.; Nagasako, M.; Chiba, A. Refinement of lamellar structures in Ti-Al alloy. *Acta Mater.* **2017**, *125*, 81–97. [CrossRef]
40. Nishikiori, S.; Takahashi, S.; Satou, S.; Tanaka, T.; Matsuo, T. Microstructure and creep strength of fully-lamellar TiAl alloys containing beta-phase. *Mater. Sci. Eng. A* **2002**, *329–331*, 802–809. [CrossRef]

41. Sufiiarov, V.S.; Popovich, A.A.; Borisov, E.V.; Polozov, I.A.; Masaylo, D.V.; Orlov, A.V. The Effect of Layer Thickness at Selective Laser Melting. *Procedia Eng.* **2017**, *174*, 126–134. [CrossRef]
42. Zhang, X.; Li, C.; Zheng, M.; Ye, Z.; Yang, X.; Gu, J. Anisotropic tensile behavior of Ti-47Al-2Cr-2Nb alloy fabricated by direct laser deposition. *Addit. Manuf.* **2020**, *32*, 101087. [CrossRef]
43. Thijs, L.; Verhaeghe, F.; Craeghs, T.; Van Humbeeck, J.; Kruth, J.-P. A study of the microstructural evolution during selective laser melting of Ti–6Al–4V. *Acta Mater.* **2010**, *58*, 3303–3312. [CrossRef]
44. Tang, H.P.; Yang, G.Y.; Jia, W.P.; He, W.W.; Lu, S.L.; Qian, M. Additive manufacturing of a high niobium-containing titanium aluminide alloy by selective electron beam melting. *Mater. Sci. Eng. A* **2015**, *636*, 103–107. [CrossRef]
45. Palm, M.; Zhang, L.; Stein, F.; Sauthoff, G. Phases and phase equilibria in the Al-rich part of the Al–Ti system above 900 °C. *Intermetallics* **2002**, *10*, 523–540. [CrossRef]
46. Draper, S.L.; Isheim, D. Environmental embrittlement of a third generation γ TiAl alloy. *Intermetallics* **2012**, *22*, 77–83. [CrossRef]
47. Zhu, H.; Seo, D.Y.; Maruyama, K. Strengthening behavior of beta phase in lamellar microstructure of TiAl alloys. *JOM* **2010**, *62*, 64–69. [CrossRef]
48. Loeber, L.; Biamino, S.; Ackelid, U.; Sabbadini, S.; Epicoco, P.; Fino, P.; Eckert, J. Comparison of selective laser and electron beam melted titanium aluminides. In Proceedings of the 22nd Annual International Solid Freeform Fabrication Symposium—An Additive Manufacturing Conference, SFF 2011, University of Texas, Austin, TX, USA, 8–10 August 2011.
49. Shazly, M.; Prakash, V.; Draper, S. Mechanical behavior of Gamma-Met PX under uniaxial loading at elevated temperatures and high strain rates. *Int. J. Solids Struct.* **2004**, *41*, 6485–6503. [CrossRef]
50. Kim, Y.-K.; Hong, J.K.; Lee, K.-A. Enhancing the creep resistance of electron beam melted gamma Ti–48Al–2Cr–2Nb alloy by using two-step heat treatment. *Intermetallics* **2020**, *121*, 106771. [CrossRef]
51. Lamirand, M.; Bonnentien, J.-L.; Guérin, S.; Ferrière, G.; Chevalier, J.-P. Effects of interstitial oxygen on microstructure and mechanical properties of Ti-48Al-2Cr-2Nb with fully lamellar and duplex microstructures. *Metall. Mater. Trans. A* **2006**, *37*, 2369–2378. [CrossRef]
52. Abiola Raji, S.; Patricia Idowu Popoola, A.; Leslie Pityana, S.; Muhmmed Popoola, O.; Olufemi Aramide, F. Laser Based Additive Manufacturing Technology for Fabrication of Titanium Aluminide-Based Composites in Aerospace Component Applications. In *Environmental Impact of Aviation and Sustainable Solutions [Working Title]*; IntechOpen: London, UK, 2019.

© 2020 by the authors. Licensee MDPI, Basel, Switzerland. This article is an open access article distributed under the terms and conditions of the Creative Commons Attribution (CC BY) license (http://creativecommons.org/licenses/by/4.0/).

Article

Microstructure and Mechanical Properties of NiTi-Based Eutectic Shape Memory Alloy Produced via Selective Laser Melting In-Situ Alloying by Nb

Igor Polozov * and Anatoly Popovich

Institute of Mechanical Engineering, Materials, and Transport, Peter the Great St. Petersburg Polytechnic University, Polytechnicheskaya 29, 195251 St. Petersburg, Russia; director@immet.spbstu.ru
* Correspondence: polozov_ia@spbstu.ru

Abstract: This paper presents the results of selective laser melting (SLM) process of a nitinol-based NiTiNb shape memory alloy. The eutectic alloy $Ni_{45}Ti_{45}Nb_{10}$ with a shape memory effect was obtained by SLM in-situ alloying using a powder mixture of NiTi and Nb powder particles. Samples with a high relative density (>99%) were obtained using optimized process parameters. Microstructure, phase composition, tensile properties, as well as martensitic phase transformations temperatures of the produced alloy were investigated in as-fabricated and heat-treated conditions. The NiTiNb alloy fabricated using the SLM in-situ alloying featured the microstructure consisting of the NiTi matrix, fine NiTi+β-Nb eutectics, as well as residual unmelted Nb particles. The mechanical tests showed that the obtained alloy has a yield strength up to 436 MPa and the tensile strength up to 706 MPa. At the same time, in-situ alloying with Nb allowed increasing the hysteresis of martensitic transformation as compared to the alloy without Nb addition from 22 to 50 °C with an increase in A_f temperature from −5 to 22 °C.

Keywords: additive manufacturing; powder bed fusion; nitinol

Citation: Polozov, I.; Popovich, A. Microstructure and Mechanical Properties of NiTi-Based Eutectic Shape Memory Alloy Produced via Selective Laser Melting In-Situ Alloying by Nb. *Materials* **2021**, *14*, 2696. https://doi.org/10.3390/ma14102696

Academic Editor: Federica Bondioli

Received: 21 April 2021
Accepted: 19 May 2021
Published: 20 May 2021

Publisher's Note: MDPI stays neutral with regard to jurisdictional claims in published maps and institutional affiliations.

Copyright: © 2021 by the authors. Licensee MDPI, Basel, Switzerland. This article is an open access article distributed under the terms and conditions of the Creative Commons Attribution (CC BY) license (https://creativecommons.org/licenses/by/4.0/).

1. Introduction

NiTi nitinol-based alloys are among the most important shape memory and superplastic materials and are of considerable interest for potential applications in aerospace and biomedical fields [1,2]. Nitinol has found its practical applications in the aircraft industry for thermo-power actuators, thermomechanical connectors, in medicine for stents, constrictive fixators, and other medical devices [3,4]. Nitinol-based alloys, unlike Fe- and Cu-based shape memory alloys, are characterized by high strength and ductility and have high corrosion resistance and biocompatibility. However, brittle secondary phases such as the metastable compound Ni_4Ti_3, which can decompose into Ni_3Ti_2 and Ni_3Ti with increasing temperature, are commonly observed in NiTi-based alloys [5,6]. These phases increase the hardness of the material, but reduce the ductility of NiTi alloys, limiting their application. In this regard, it is relevant to introduce secondary phases with high hardness into the matrix of NiTi-based alloys while maintaining or slightly reducing the level of ductility.

It is known that Nb plays an important role in nitinol-based alloys because the addition of Nb increases the hysteresis of martensitic transformation [7]. Besides, the addition of Nb improves the biocompatibility of NiTi-based alloys [8]. Extension of the temperature interval of martensitic transformation can be used in the manufacture of connecting elements or seals [9]. $Ni_{47}Ti_{44}Nb_9$ alloy is a classic nitinol-based shape memory alloy characterized by an in situ composite microstructure [7]. It is known that the microstructure of the alloy with shape memory effect $Ni_{47}Ti_{44}Nb_9$ consists of NiTi matrix and β-Nb phase [10]. The presence of β-Nb is known to have a significant effect on the shape memory effect and mechanical properties of the alloy [11]. Nb exhibits different properties from the NiTi matrix with a shape memory effect [12]. The presence of Nb as a second phase (β-Nb) or as a solution is the primary reason for transformation hysteresis expansion in NiTiNb shape memory alloys [10].

Conventionally, $Ni_{47}Ti_{44}Nb_9$ alloy parts are made by casting or sintering powder materials, which have limitations in terms of part geometry, as well as control of their microstructure and properties [13–15]. In this regard, it is promising to use additive manufacturing (AM) techniques, in particular, the selective laser melting (SLM) method for the manufacture of parts from alloys with the shape memory effect [16].

The SLM process uses metallic powders as a feedstock material. Various research works have shown that pre-alloyed powders allow obtaining a material with a more homogeneous microstructure and stable mechanical properties [17–19]. However, the production of pre-alloyed powders is usually time and labor consuming, especially in the case of custom alloys. The use of an elemental powder blend in the SLM process is an alternative option that can be used for in-situ synthesis or in-situ alloying during the SLM process to obtain a material with a required composition [20–22]. At the same time, rapid solidification and cooling rates typical for the SLM process lead to insufficient diffusion of elements with high melting points, which causes a heterogeneous microstructure. In this regard, additional heat treatment can be applied to improve the chemical homogeneity of the material [23,24]. AM methods, such as SLM or Direct Energy Deposition (DED), involve repetitive heating and high solidification rates leading to distinctive microstructural features of nitinol alloys [25]. For example, rapid solidification promotes the formation of a supersaturated solid solution matrix [25]. At the same time, various process parameters can result in different microstructures and phase composition of nitinol alloys [26], as well as different Ni content resulting in significant changes in transformation behavior [16].

In this work, the feasibility of the SLM process to produce a nitinol-based shape memory alloy via in-situ alloying by Nb was investigated. Using a powder blend of NiTi-alloy and pure Nb, the influence of the SLM process parameters on the sample's density was studied. Microstructure, phase composition, mechanical properties, and martensitic phase transformation temperatures were studied for the NiTiNb alloy in the as-fabricated and heat-treated conditions.

2. Materials and Methods

The following materials were used as feedstock materials: gas atomized NiTi alloy powder with a nickel content of 51.4% (at.) and niobium powder with a purity of 99.8%. The NiTi powder had the following particle size distribution: d_{10} = 27.2 µm, d_{50} = 50.0 µm, and d_{90} = 84.9 µm. The niobium powder in the initial state had a spherical particle shape and was pretreated in a thermal plasma jet using Tekna Tek-15 (Sherbrooke, QC, Canada) plasma spheroidization unit to obtain spherical particles. The details of the plasma spheroidization process can be found in [27]. The final niobium powder had the following particle size distribution: d_{10} = 15.5 µm, d_{50} = 35.5 µm, and d_{90} = 72.8 µm. Particle size distribution of the powders was measured by laser diffraction technique with Analysette 22 NanoTec (Fritsch, Idar-Oberstein, Germany).

NiTi and Nb powders were mixed in 85% NiTi: 15% Nb weight ratio for 12 h using a tumbler mixer to obtain a powder blend with $Ni_{47}Ti_{44}Nb_9$ composition. Figure 1a shows a scanning electron microscope (SEM) image of the NiTi and Nb powder blend. The NiTi powder particles have a spherical shape, while some Nb particles have irregular shape due to incomplete spheroidization. Figure 1b,c shows the distribution of Ti, Ni, and Nb elements in the powder blend demonstrating that Nb particles were fairly distributed among the NiTi powder.

The NiTiNb alloy samples were manufactured from the prepared powder blend using AconityMIDI (Aconity3D GmbH, Herzogenrath, Germany) SLM system with varying process parameters. Samples with the size of 10 × 10 × 10 mm³ were produced to study the microstructure, phase composition, and density of the material. Two groups of samples were fabricated, the difference between which is the use of different powder layer thickness and the laser beam spot size. Laser power, scanning speed, and hatch distance were also varied. The SLM process parameters used in the study are shown in Table 1. Sets A, B, and C use standard laser spot size (~70 µm) and sets G, K, and J use increased laser spot size with an unfocused beam (~300 µm). The volume energy density (VED) calculated according to a standard equation [28,29] was used as a parameter to investigate the effects of SLM process parameters on the samples' density.

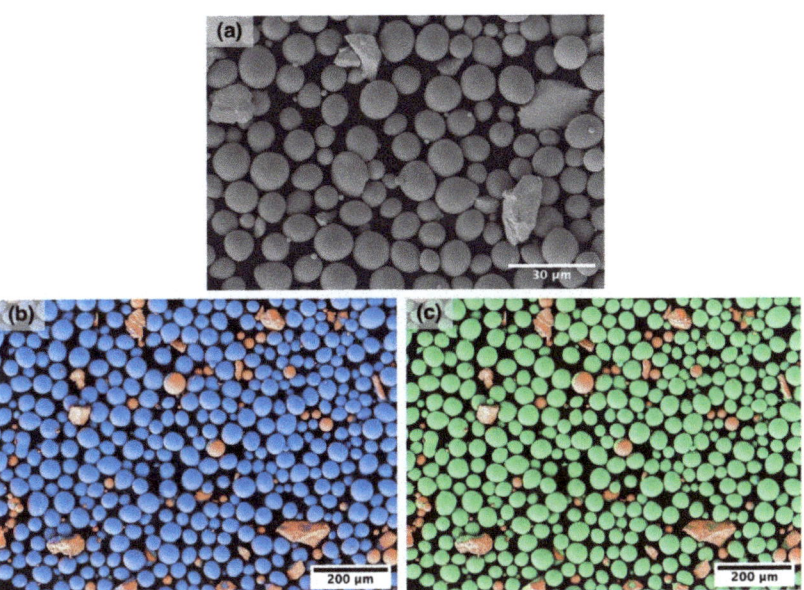

Figure 1. (**a**) SEM-image of the NiTi-Nb powder blend and chemical distribution of the elements in the blend: (**b**) blue—Ti and red—Nb and (**c**) green—Ni and red—Nb.

Table 1. SLM process parameters used to fabricate the samples.

Process Parameter Set	Power (W)	Scanning Speed (mm/s)	Hatch Distance (µm)	Layer Thickness (µm)	VED (J/mm^3)
A1	200	600	120		55.6
A2	200	650	120		51.3
A3	200	700	120		47.6
B1	180	600	120	50	50.0
B2	240	600	120		66.7
C1	200	600	100		66.7
C2	200	600	90		74.1
D1	450	300	450		33.3
D2	450	320	450		31.3
D3	450	340	450		29.4
E1	400	340	450	100	26.1
E2	420	340	450		27.5
F1	450	340	400		33.1
F2	450	340	350		37.8

To study the effect of heat treatment temperature and holding time on the microstructure and properties of the samples, annealing was carried out under the following conditions:

- heating to 500 °C and holding at 500 °C for 2 h;
- heating to 900 °C and holding at 900 °C for 30 min;
- heating to 900 °C and holding at 900 °C for 2 h.

The samples were heated at a rate of 10 °C/min and furnace cooled. The annealing temperature of 900 °C is above the recrystallization temperature of NiTiNb alloys [30], while 500 °C is below the recrystallization temperature and its main purpose is residual stress relieving.

The microstructure was studied using a TESCAN Mira 3 LMU scanning electron microscope (SEM) (TESCAN, Brno, Czech Republic) and a Leica DMI 5000 optical microscope (Leica, Wetzlar, Germany). To study the microstructure, polished microsections of the samples were etched using the following etchant: 83% H_2O, 14% HNO_3, and 3% HF. The samples were cut along the building direction. The chemical composition of the material was investigated using energy-dispersive analysis (EDS) (TESCAN, Brno, Czech Republic). The phase composition of the material was determined using a Bruker D8 Advance diffractometer on CuKα radiation (λ = 1.5418 Å). The density of the material was measured by the Archimedes principle. Tensile mechanical properties were investigated using cylindrical specimens using a Zwick/Roell Z050 testing machine (Ulm, Germany). Temperatures of martensitic transformations of the fabricated alloy were determined by differential scanning calorimetry (DSC). To compare phase transformation temperatures, NiTi samples were fabricated without the addition of Nb particles using the E2 parameter set.

3. Results and Discussion

Figure 2 shows the effect of VED on the density of NiTiNb samples produced using the NiTi-Nb powder blend with 300 µm (Figure 2a) and 70 µm (Figure 2b) laser spot size. The highest relative density values of around 99.2 ± 0.05% were obtained using E1 and E2 parameter sets, which correspond to 300 µm laser spot size and 26–27 J/mm³ VED. Increasing VED led to lower density values, which might be attributed to melt pool overheating and instability and formation of gas and keyhole pores [31,32]. At the same time, varying VED at 70 µm laser spot size and 50 µm layer thickness did not lead to a significant change in density.

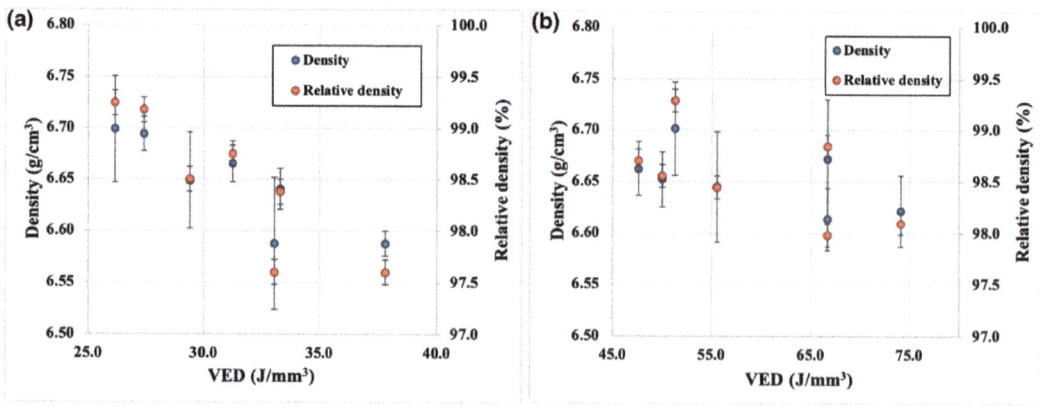

Figure 2. Effect of VED on the density of NiTiNb samples produced by SLM using (**a**) 300 µm and (**b**) 70 µm laser spot size.

As can be seen in Figure 3, there are several types of internal defects in samples produced under different process parameters. In the case of the samples D3 and F1, coarse pores with the size of 300–400 µm and irregular shape can be found along with fine spherical pores. These coarse pores elongated in the direction parallel to the laser path might be the result of lack-of-fusion due to the interaction of various factors: capillary forces, material evaporation, insufficient melting, etc. [33,34]. The smallest number of defects can be seen in the case of the sample E2, which corresponds to 27.5 J/mm^3 VED and 300 µm laser spot and has the highest density value.

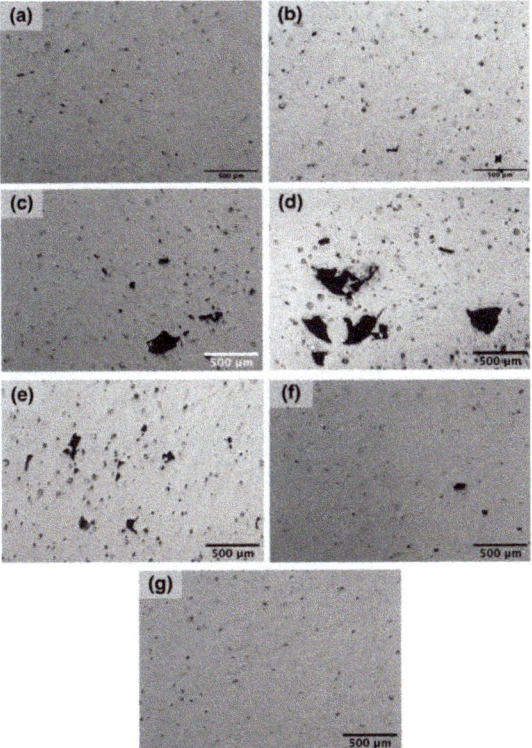

Figure 3. Optical images of the polished microsections for the samples produced using parameter sets (**a**) D1, (**b**) D2, (**c**) D3, (**d**) F1, (**e**) F2, (**f**) E1, and (**g**) E2.

Figure 4 shows SEM-images of microstructures of the samples manufactured using E1 and A1 parameter sets. In the case of the E1 sample produced with an unfocused laser beam and 100 µm layer thickness, there are coarse solidified melt pools with a width comparable to the diameter of the laser spot as can be seen in Figure 4a. Melt pool boundaries can be clearly distinguished due to the presence of eutectic bands.

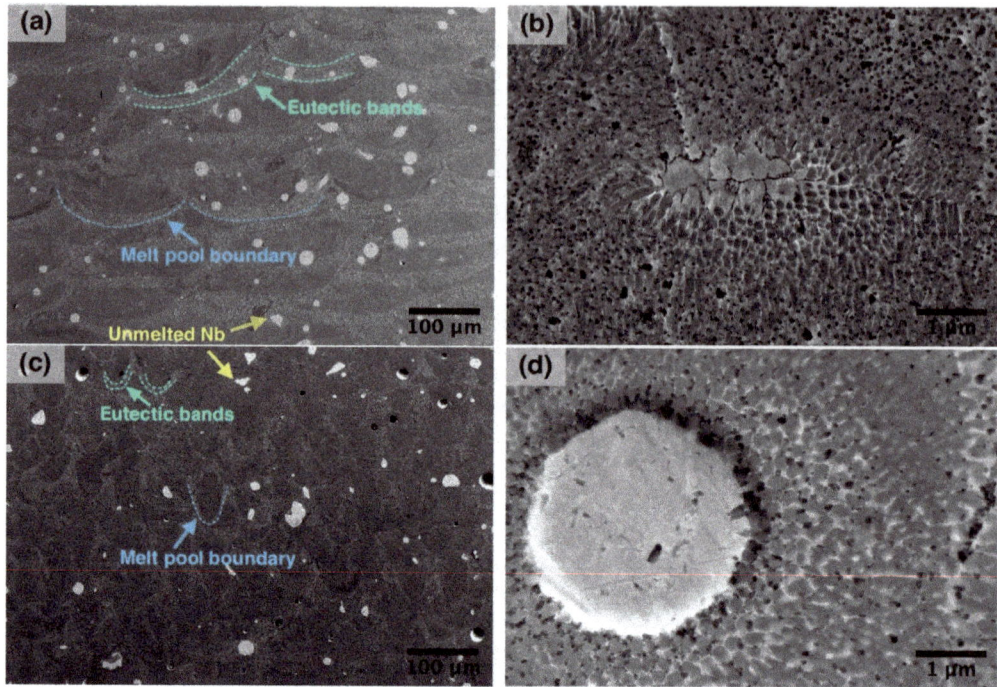

Figure 4. SEM-images showing the microstructure of the NiTiNb samples produced using parameter sets (**a,b**) E1 and (**c,d**) A1.

TiNi-Nb alloy system is a eutectic alloy and according to the phase diagram, the eutectic point corresponds to 26 at. % Ni content [35]. The eutectic temperature is 1150.7 °C. Hence, NiTi+β-Nb eutectic should be present in the microstructure after solidification and cooling of the Ti-Ni-Nb alloy [36]. The presence of NiTi and β-Nb phase was confirmed by the XRD results (Figure 5). The eutectic is mainly located at the melt pool boundaries as can be seen in Figure 4. Nb has a significantly higher density and melting point compared to NiTi. As a consequence, during the SLM of the powder blend, melting of NiTi will occur first, while the Nb may remain unmelted. Mixing of the alloy components in the melt can be carried out by Marangoni convection [37], which would result in a distribution of Nb in the volume of a melt pool. During the laser melting of the powder layer, the underlying solidified material is partially remelted. This can promote the diffusion at the melt pool boundaries and the dissolution of the elements in these micro volumes. Thus, the eutectic NiTI+β-Nb microstructure can be mainly found at the melt pool boundaries. As can be seen in Figure 4b,d, NiTi+β-eutectic areas are visible near partially melted on unmelted Nb particles where diffusion takes place during the SLM process and promotes the formation of fine eutectic microstructure.

When 50 μm layer thickness and a standard laser spot size were applied during the SLM process (sample A1), the microstructure of the obtained alloy also features melt pool boundaries as can be seen in Figure 4c, but their size is much smaller compared to E1 sample. The eutectic bands width is also smaller in this case, which indicates less degree of Nb diffusion into the NiTi matrix in case of smaller laser spot size. When an unfocused laser beam along with increased laser power and layer thickness is used, the melt pool volume is bigger and a higher volume of powder blend is melted with the laser. At the same time, a bigger melt pool volume leads to a lower cooling rate during the SLM process [38], which promotes the diffusion of elements and formation of the eutectic phase. Thus, a higher volume of eutectic phase is obtained when an increased laser spot size is used.

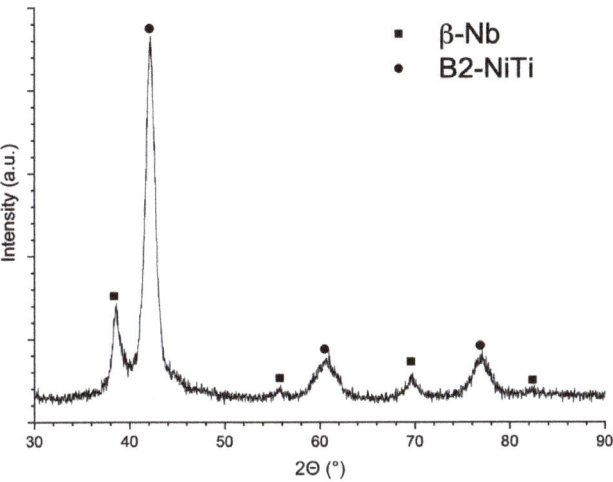

Figure 5. The XRD pattern of the NiTiNb sample produced using the E1 parameter set.

Figure 6 shows the chemical distribution of the elements in the NiTiNb sample produced using the E1 parameter set. It can be seen that there are areas of pure Nb corresponding to unmelted Nb particles. In general, Ti, Ni, and Nb are homogeneously distributed in the volume, and the eutectic regions do not feature a significant change in concentration of the elements compared to the rest of the sample. According to the EDS results, the obtain alloy has the following composition (in at. %): 45.5 ± 0.2 Ti, 44.9 ± 0.2 Ni, and 9.6 ± 0.4 Nb. The obtained composition is characterized by a decreased Ni content due to its evaporation during the SLM process, which is consistent with the results on the SLM of NiTi alloys reported in the literature [16,39].

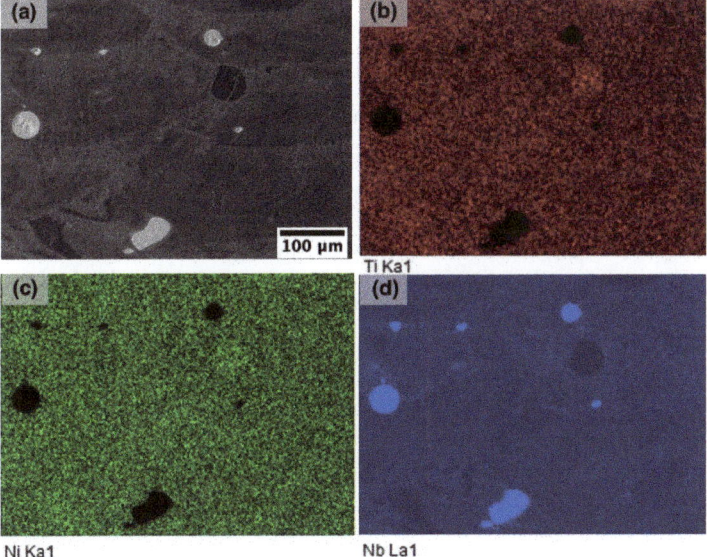

Figure 6. (**a**) SEM-image of the microstructure of E1 sample and EDS-maps showing the distribution of (**b**) Ti, (**c**) Ni, and (**d**) Nb.

To obtain a more homogeneous structure, the samples manufactured using the E1 parameter set were heat-treated under different conditions. SEM-images of the microstructures after heat treatment are shown in Figure 7. As a result of heat treatment, Nb diffusion into the material matrix occurs, which is observed in the form of partial dissolution of Nb particles and an increase in the eutectic volume fraction, but the annealing temperature and time are not sufficient for a complete dissolution of Nb, because it has a low diffusion coefficient. Thus, after annealing at 900 °C for 30 min, the alloy structure has not undergone significant changes. As the annealing time was increased to 2 h at 900 °C, the former melt pool boundaries became less pronounced, and the fraction of regions with eutectic microstructure increased due to an accelerated niobium diffusion in the alloy matrix.

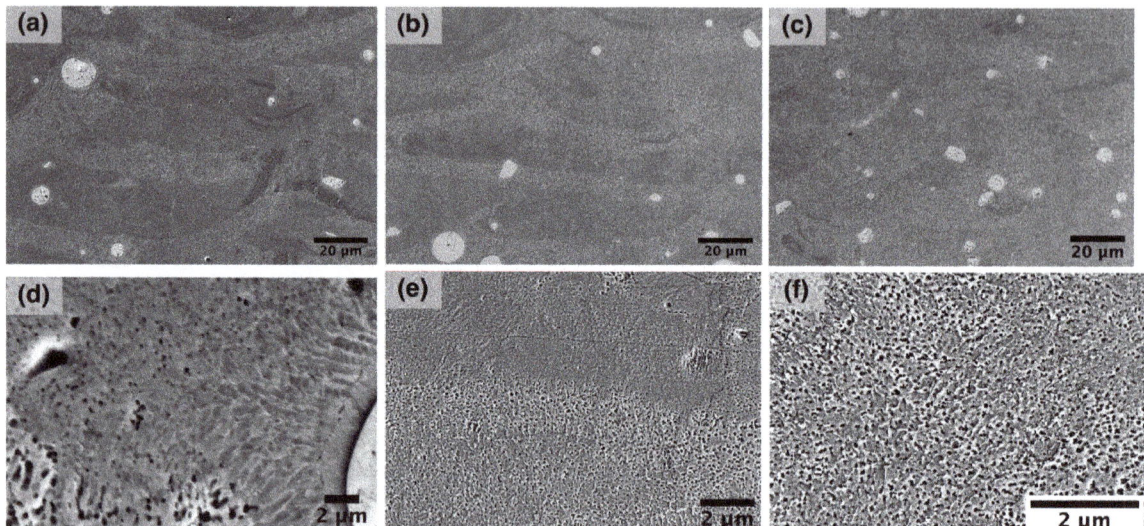

Figure 7. SEM-images of the NiTiNb samples after different heat treatments: (**a,d**) 500 °C for 2 h, (**b,e**) 900 °C for 30 min, and (**c,f**) 900 °C for 2 h.

For the tensile tests, the samples were heat-treated by annealing at 900 °C for 2 h since it provides a more homogeneous microstructure of the alloy. The results of the tensile tests are shown in Table 2. While the tensile strength of the in-situ alloyed samples is comparable to the casted $Ni_{47}Ti_{44}Nb_9$ alloy, the elongation is significantly lower. Heterogeneous microstructure and impurities pick-up are believed to be the main factors affecting the elongation of the SLM-ed alloy.

Table 2. The results of tensile tests at room temperature for NiTiNb samples in the as-fabricated and heat-treated state.

Material	Yield Strength (MPa)	Tensile Strength (MPa)	Elongation (%)
NiTi-+Nb, SLM, as-fabricated	390 ± 10	590 ± 60	1.5 ± 0.1
NiTi-+Nb, SLM, after H/T (900 °C for 2 h)	410 ± 20	680 ± 20	3.8 ± 0.3
$Ni_{47}Ti_{44}Nb_9$, casted [10]	~500	~650	~40

Figure 8 shows the DSC curves for the fabricated samples. In the as-fabricated condition, the reverse martensitic transformation effect is poorly visible. However, the presence of a reverse martensitic transformation peak allows for the conclusion that a forward martensitic transformation also takes place during cooling. Annealing at 900 °C resulted in A_f temperature shift into higher temperature range and made both forward and reverse transformation effects more pronounced. Table 3 summarizes the DSC results of the obtained samples showing martensitic phase transformation temperatures in the as-fabricated and heat-treated states. One of the main functional properties of shape memory alloys are temperatures of direct and reverse martensitic phase transformations. It is reasonable to consider that the solution of Nb in the NiTi matrix can change the kinetics of martensitic transformations [10].

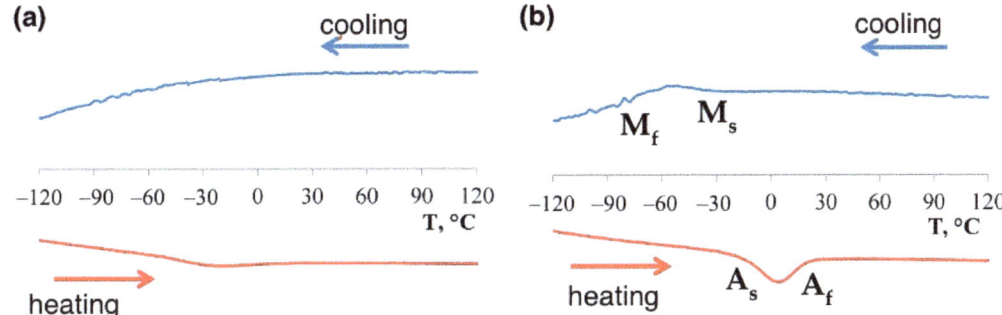

Figure 8. DSC-curves of the in-situ alloyed NiTiNb samples (**a**) in the as-fabricated state and (**b**) after annealing at 900 °C for 2 h.

Table 3. Martensitic phase transformation temperatures of the fabricated alloys.

Sample	M_s (°C)	M_f (°C)	A_s (°C)	A_f (°C)	A_f-A_s (°C)
NiTi+Nb, SLM as-fabricated	−50	−78	−52	5	57
NiTi+Nb, SLM+H/T (900 °C for 2 h)	−30	−76	−28	22	50
NiTi, SLM as-fabricated	−34	−69	−27	−5	22
$Ni_{47}Ti_{44}Nb_9$, casted [40]	−73	−90	−25	−11	14

Annealing at 900 °C leads to increased and narrowed temperature intervals of the reverse martensitic transformation compared to the as-fabricated state. The A_f temperature shifts to the region of positive temperatures. At the same time, "in-situ" alloying increased the hysteresis of martensitic transformation from 22 to 50 °C as compared to NiTi alloy without Nb addition with an increase in the A_f temperature from −5 to 22 °C. The in-situ alloyed NiTiNb samples showed higher direct martensitic transformation temperatures compared to the casted $Ni_{47}Ti_{44}Nb_9$ alloy, which might be attributed to the difference in Ni content. Due to partial evaporation, the SLM-ed samples have a Ni content of around 45.4 at. % leading to the increase of transformation temperatures.

4. Conclusions

The SLM process of NiTi-based eutectic alloy with shape memory effect obtained by in-situ alloying with Nb has been studied. Based on the results obtained, the following main conclusions can be made:

- NiTiNb shape memory alloy can be produced by SLM in-situ alloying by Nb with a relative density of 99%.
- The microstructure of the in-situ alloyed material consists of B2-NiTi matrix, fine NiTi + β-Nb eutectic phase, and residual unmelted Nb particles.

- The use of increased laser spot size with simultaneous increase of layer thickness and laser power allows to obtain more homogeneous element distribution and homogeneous microstructure of NiTiNb alloy.
- The SLM in-situ alloying of NiTi by Nb allowed increasing the martensitic transformation hysteresis as compared to the alloy without Nb addition from 22 to 50 °C while the A_f temperature increased from −5 to 22 °C.
- Annealing of the in-situ alloyed material at 900 °C resulted in improved microstructural homogeneity and higher tensile strength.

Author Contributions: Conceptualization, I.P.; investigation, I.P.; methodology, I.P.; project administration, A.P.; supervision, A.P.; writing—original draft, I.P.; writing—review and editing, A.P. All authors have read and agreed to the published version of the manuscript.

Funding: This study was carried out under the State Contract dated 4 June 2020, No. H.4ш.241.09.20.108 (ISC 17706413348200001110).

Institutional Review Board Statement: Not applicable.

Informed Consent Statement: Not applicable.

Data Availability Statement: The data presented in this study are available on request from the corresponding author.

Conflicts of Interest: The authors declare no conflict of interest.

References

1. Otsuka, K.; Ren, X. Physical metallurgy of Ti–Ni-based shape memory alloys. *Prog. Mater. Sci.* **2005**, *50*, 511–678. [CrossRef]
2. Lee, J.-H.; Lee, K.-J.; Choi, E. Flexural capacity and crack-closing performance of NiTi and NiTiNb shape-memory alloy fibers randomly distributed in mortar beams. *Compos. Part B Eng.* **2018**, *153*, 264–276. [CrossRef]
3. Farber, E.; Zhu, J.-N.; Popovich, A.; Popovich, V. A review of NiTi shape memory alloy as a smart material produced by additive manufacturing. *Mater. Today Proc.* **2020**, *30*, 761–767. [CrossRef]
4. Mohd Jani, J.; Leary, M.; Subic, A.; Gibson, M.A. A review of shape memory alloy research, applications and opportunities. *Mater. Des.* **2014**, *56*, 1078–1113. [CrossRef]
5. Wang, L.; Wang, C.; Zhang, L.-C.; Chen, L.; Lu, W.; Zhang, D. Phase transformation and deformation behavior of NiTi-Nb eutectic joined NiTi wires. *Sci. Rep.* **2016**, *6*, 23905. [CrossRef]
6. Khanlari, K.; Shi, Q.; Li, K.; Xu, P.; Cao, P.; Liu, X. An investigation into the possibility to eliminate the microstructural defects of parts printed using a Ni-rich Ni-Ti elemental powder mixture. *Mater. Res. Express* **2020**, *7*, 106503. [CrossRef]
7. BAO, Z.; GUO, S.; XIAO, F.; ZHAO, X. Development of NiTiNb in-situ composite with high damping capacity and high yield strength. *Prog. Nat. Sci. Mater. Int.* **2011**, *21*, 293–300. [CrossRef]
8. Zhang, L.-C.; Chen, L.-Y.; Wang, L. Surface Modification of Titanium and Titanium Alloys: Technologies, Developments, and Future Interests. *Adv. Eng. Mater.* **2020**, *22*, 1901258. [CrossRef]
9. Dong, Z.Z.; Zhou, S.L.; Liu, W.X. A Study of NiTiNb Shape-Memory Alloy Pipe-Joint with Improved Properties. *Mater. Sci. Forum* **2002**, *394–395*, 107–110. [CrossRef]
10. Wei, L.; Xinqing, Z. Mechanical Properties and Transformation Behavior of NiTiNb Shape Memory Alloys. *Chinese J. Aeronaut.* **2009**, *22*, 540–543. [CrossRef]
11. Ying, C.; Hai-chang, J.; Li-jian, R.; Li, X.; Xin-qing, Z. Mechanical behavior in NiTiNb shape memory alloys with low Nb content. *Intermetallics* **2011**, *19*, 217–220. [CrossRef]
12. Fu, X.; Guojun, M.; Xinqing, Z.; Huibin, X. Effects of Nb Content on Yield Strength of NiTiNb Alloys in Martensite State. *Chinese J. Aeronaut.* **2009**, *22*, 658–662. [CrossRef]
13. Taylor, S.L.; Ibeh, A.J.; Jakus, A.E.; Shah, R.N.; Dunand, D.C. NiTi-Nb micro-trusses fabricated via extrusion-based 3D-printing of powders and transient-liquid-phase sintering. *Acta Biomater.* **2018**, *76*, 359–370. [CrossRef] [PubMed]
14. Wang, Y.; Jiang, S.; Zhang, Y. Processing Map of NiTiNb Shape Memory Alloy Subjected to Plastic Deformation at High Temperatures. *Metals* **2017**, *7*, 328. [CrossRef]
15. Slipchenko, V.N.; Koval, Y.N.; Koshovy, O.V. Influence of casting technology on the phase transformation in NiTiNb alloys. *J. Phys. IV* **2003**, *112*, 717–719. [CrossRef]
16. Wang, X.; Yu, J.; Liu, J.; Chen, L.; Yang, Q.; Wei, H.; Sun, J.; Wang, Z.; Zhang, Z.; Zhao, G.; et al. Effect of process parameters on the phase transformation behavior and tensile properties of NiTi shape memory alloys fabricated by selective laser melting. *Addit. Manuf.* **2020**, *36*, 101545. [CrossRef]

17. Polozov, I.; Sufiiarov, V.; Kantyukov, A.; Razumov, N.; Goncharov, I.; Makhmutov, T.; Silin, A.; Kim, A.; Starikov, K.; Shamshurin, A.; et al. Microstructure, densification, and mechanical properties of titanium intermetallic alloy manufactured by laser powder bed fusion additive manufacturing with high-temperature preheating using gas atomized and mechanically alloyed plasma spheroidized powders. *Addit. Manuf.* **2020**, *34*, 101374. [CrossRef]
18. Grigoriev, A.; Polozov, I.; Sufiiarov, V.; Popovich, A. In-situ synthesis of Ti2AlNb-based intermetallic alloy by selective laser melting. *J. Alloys Compd.* **2017**, *704*, 434–442. [CrossRef]
19. Fischer, M.; Joguet, D.; Robin, G.; Peltier, L.; Laheurte, P. In situ elaboration of a binary Ti–26Nb alloy by selective laser melting of elemental titanium and niobium mixed powders. *Mater. Sci. Eng. C* **2016**, *62*, 852–859. [CrossRef] [PubMed]
20. Simonelli, M.; Aboulkhair, N.T.; Cohen, P.; Murray, J.W.; Clare, A.T.; Tuck, C.; Hague, R.J.M. A comparison of Ti-6Al-4V in-situ alloying in Selective Laser Melting using simply-mixed and satellited powder blend feedstocks. *Mater. Charact.* **2018**, *143*, 118–126. [CrossRef]
21. Hanemann, T.; Carter, L.N.; Habschied, M.; Adkins, N.J.E.; Attallah, M.M.; Heilmaier, M. In-situ alloying of AlSi10Mg+Si using Selective Laser Melting to control the coefficient of thermal expansion. *J. Alloys Compd.* **2019**, *795*, 8–18. [CrossRef]
22. Katz-Demyanetz, A.; Koptyug, A.; Popov, V.V. In-situ Alloying as a Novel Methodology in Additive Manufacturing. In Proceedings of the 2020 IEEE 10th International Conference Nanomaterials: Applications & Properties (NAP), Sumy, Ukraine, 9–13 November 2020; pp. 02SAMA05-1–02SAMA05-4.
23. Polozov, I.; Sufiiarov, V.; Kantyukov, A.; Popovich, A. Selective Laser Melting of Ti2AlNb-based intermetallic alloy using elemental powders: Effect of process parameters and post-treatment on microstructure, composition, and properties. *Intermetallics* **2019**, *112*, 106554. [CrossRef]
24. Wang, J.C.; Liu, Y.J.; Qin, P.; Liang, S.X.; Sercombe, T.B.; Zhang, L.C. Selective laser melting of Ti–35Nb composite from elemental powder mixture: Microstructure, mechanical behavior and corrosion behavior. *Mater. Sci. Eng. A* **2019**, *760*, 214–224. [CrossRef]
25. Zhang, Q.; Hao, S.; Liu, Y.; Xiong, Z.; Guo, W.; Yang, Y.; Ren, Y.; Cui, L.; Ren, L.; Zhang, Z. The microstructure of a selective laser melting (SLM)-fabricated NiTi shape memory alloy with superior tensile property and shape memory recoverability. *Appl. Mater. Today* **2020**, *19*, 100547. [CrossRef]
26. Gan, J.; Duan, L.; Li, F.; Che, Y.; Zhou, Y.; Wen, S.; Yan, C. Effect of laser energy density on the evolution of Ni4Ti3 precipitate and property of NiTi shape memory alloys prepared by selective laser melting. *J. Alloys Compd.* **2021**, *869*, 159338. [CrossRef]
27. Goncharov, I.S.; Masaylo, D.V.; Orlov, A.; Razumov, N.G.; Obrosov, A. The Effect of Laser Power on the Microstructure of the Nb-Si Based In Situ Composite, Fabricated by Laser Metal Deposition. *Key Eng. Mater.* **2019**, *822*, 556–562. [CrossRef]
28. Enneti, R.K.; Morgan, R.; Atre, S.V. Effect of process parameters on the Selective Laser Melting (SLM) of tungsten. *Int. J. Refract. Met. Hard Mater.* **2018**, *71*, 315–319. [CrossRef]
29. Nandwana, P.; Elliott, A.M.; Siddel, D.; Merriman, A.; Peter, W.H.; Babu, S.S. Powder bed binder jet 3D printing of Inconel 718: Densification, microstructural evolution and challenges. *Curr. Opin. Solid State Mater. Sci.* **2017**, *21*, 207–218. [CrossRef]
30. Suhail, R.; Amato, G.; McCrum, D. Heat-activated prestressing of NiTiNb shape memory alloy wires. *Eng. Struct.* **2020**, *206*, 110128. [CrossRef]
31. Kasperovich, G.; Haubrich, J.; Gussone, J.; Requena, G. Correlation between porosity and processing parameters in TiAl6V4 produced by selective laser melting. *Mater. Des.* **2016**, *105*, 160–170. [CrossRef]
32. Yadroitsev, I.; Krakhmalev, P.; Yadroitsava, I. Hierarchical design principles of selective laser melting for high quality metallic objects. *Addit. Manuf.* **2014**, *7*, 45–56. [CrossRef]
33. Bayat, M.; Mohanty, S.; Hattel, J.H. Multiphysics modelling of lack-of-fusion voids formation and evolution in IN718 made by multi-track/multi-layer L-PBF. *Int. J. Heat Mass Transf.* **2019**, *139*, 95–114. [CrossRef]
34. Martin, A.A.; Calta, N.P.; Khairallah, S.A.; Wang, J.; Depond, P.J.; Fong, A.Y.; Thampy, V.; Guss, G.M.; Kiss, A.M.; Stone, K.H.; et al. Dynamics of pore formation during laser powder bed fusion additive manufacturing. *Nat. Commun.* **2019**, *10*, 1987. [CrossRef]
35. Piao, M.; Miyazaki, S.; Otsuka, K.; Nishida, N. Effects of Nb Addition on the Microstructure of Ti–Ni Alloys. *Mater. Trans. JIM* **1992**, *33*, 337–345.
36. Fan, Q.C.; Zhang, Y.; Zhang, Y.H.; Wang, Y.Y.; Yan, E.H.; Huang, S.K.; Wen, Y.H. Influence of Ni/Ti ratio and Nb addition on martensite transformation behavior of NiTiNb alloys. *J. Alloys Compd.* **2019**, *790*, 1167–1176. [CrossRef]
37. Tan, C.; Zhou, K.; Kuang, T. Selective laser melting of tungsten-copper functionally graded material. *Mater. Lett.* **2019**, *237*, 328–331. [CrossRef]
38. Sufiiarov, V.S.; Popovich, A.A.; Borisov, E.V.; Polozov, I.A.; Masaylo, D.V.; Orlov, A.V. The Effect of Layer Thickness at Selective Laser Melting. *Procedia Eng.* **2017**, *174*, 126–134. [CrossRef]
39. Wang, Y.; Tan, X.P.; Du, Z.; Chandra, S.; Sun, Z.; Lim, C.W.J.; Tor, S.B.; Lim, C.S.; Wong, C.H. Additive manufacturing of NiTi shape memory alloys using pre-mixed powders. *J. Mater. Process. Technol.* **2019**, *271*, 152–161. [CrossRef]
40. Wang, M.; Jiang, M.; Liao, G.; Guo, S.; Zhao, X. Martensitic transformation involved mechanical behaviors and wide hysteresis of NiTiNb shape memory alloys. *Prog. Nat. Sci. Mater. Int.* **2012**, *22*, 130–138. [CrossRef]

Structure and Properties of Ti/Ti64 Graded Material Manufactured by Laser Powder Bed Fusion

Evgenii Borisov, Igor Polozov *, Kirill Starikov, Anatoly Popovich and Vadim Sufiiarov

Institute of Machinery, Materials, and Transport, Peter the Great St. Petersburg Polytechnic University (SPbPU), Polytechnicheskaya 29, 195251 St. Petersburg, Russia; evgenii.borisov@icloud.com (E.B.); kirill.starikov@gmail.com (K.S.); director@immet.spbstu.ru (A.P.); vadim.spbstu@yandex.ru (V.S.)
* Correspondence: polozov_ia@spbstu.ru

Abstract: Multimaterial additive manufacturing is an attractive way of producing parts with improved functional properties by combining materials with different properties within a single part. Pure Ti provides a high ductility and an improved corrosion resistance, while the Ti64 alloy has a higher strength. The combination of these alloys within a single part using additive manufacturing can be used to produce advanced multimaterial components. This work explores the multimaterial Laser Powder Bed Fusion (L-PBF) of Ti/Ti64 graded material. The microstructure and mechanical properties of Ti/Ti64-graded samples fabricated by L-PBF with different geometries of the graded zones, as well as different effects of heat treatment and hot isostatic pressing on the microstructure of the bimetallic Ti/Ti64 samples, were investigated. The transition zone microstructure has a distinct character and does not undergo significant changes during heat treatment and hot isostatic pressing. The tensile tests of Ti/Ti64 samples showed that when the Ti64 zones were located along the sample, the ratio of cross-sections has a greater influence on the mechanical properties than their shape and location. The presented results of the investigation of the graded Ti/Ti64 samples allow tailoring properties for the possible applications of multimaterial parts.

Keywords: additive manufacturing; selective laser melting; titanium alloys; multimaterial 3D printing; graded materials

1. Introduction

With the advent of Additive Manufacturing (AM) technologies, it has become possible for designers to improve the technological and functional capabilities of parts by evolutionary design optimization [1,2]. AM technologies have simplified the manufacturing of complex, single-piece products, while opening up the possibility of shaping a specific, given structure [3–5]. One method of part optimization is the use of multiple materials in the fabrication of a single part [6–8]. For example, in a part that is only partially exposed to high temperatures, it is possible to use heat-resistant materials only in the temperature-loaded part. In this case, for the formation of the remaining volume of the part it is reasonable to use less heat-resistant and, at the same time, cheaper materials. In addition, the combination of strong and ductile materials is widely used, for example, in tools for machining and gears, etc. [9,10]. In implants, a very important parameter is the mechanical strength and elasticity of the material. On the one hand, it is necessary to ensure a sufficient strength to avoid fracture. On the other hand, too much elasticity of the material can lead to bone damage due to permanent differences in the strain under load [11].

Recently, an increasing number of studies have appeared in the field of forming parts from several materials during the Laser Powder Bed Fusion (L-PBF) process. The main difficulty of this process is related to the fact that the existing equipment is not designed to use more than one powder material simultaneously. Therefore, much research on the development and modification of the equipment is being carried out [12–16]. For the developers, the main difficulty to overcome is the necessity to apply a thin layer of powder material of the heterogeneous chemical composition. At the same time, this heterogeneity must correspond to the computer model of the part on each layer.

Another important direction of research is the study of microstructures and the properties of the multi-material products themselves, which are obtained using the L-PBF [17–20]. In these works, the microstructure and continuity of the transition zone, its phase composition, and mechanical characteristics were investigated.

Currently, there are several research papers devoted to the L-PBF of parts with a graded composition by changing the feedstock powder to build separate parts of a specimen. For example, alternatively using powders of different compositions was applied to fabricate CuSn/18Ni300 [18], NiTi/Ti6Al4V [19], AlSi10Mg/Cr18Ni10Ti stainless steel [21], 316L/CuSn10 [22], or 316L/Cu [23] graded specimens by L-PBF. The authors of [20] investigated the possibility of using the L-PBF process to fabricate graded samples using In718 and Ti6Al4V powders utilizing intermediate layers of mixed powders with a different ratio. The first commercial multimaterial recoating system for the L-PBF machines was recently introduced by Aerosint [24,25] suggesting the importance of multimaterial AM development. It uses mechanical forces to hold the powder on the drum and can release it at the desired location, generating a 2D single material image in a line-by-line manner. Currently, its possibilities have been demonstrated by 3D printing a copper alloy/steel bi-metallic parts [26].

For medical applications, Ti and Ti64 alloys have their advantages and disadvantages. The Ti64 alloy has great strength, but pure Ti has a great resistance to corrosion [27] and does not contain toxic impurities (Vanadium). Therefore, the formation of a graded part, where the advantages of both alloys are used, is relevant.

This work aimed to investigate the microstructure and mechanical properties of Ti/Ti64 graded samples fabricated by L-PBF with different geometries of the graded zones, as well as effects of heat treatment and hot isostatic pressing on the microstructure of the bimetallic Ti/Ti64 samples.

2. Materials and Methods

Commercially available CP-Ti (grade 2) and Ti-6Al-4V (Ti64, grade 5) alloy powders (Normin LLC, Borovichi, Russia) obtained by plasma atomization process were used as the feedstock material to fabricate the samples. The particles of both powders were spherical shaped (Figure 1) and had a mean size of d_{50} = 34 μm and d_{50} = 47 μm for Ti and Ti64 alloys, respectively.

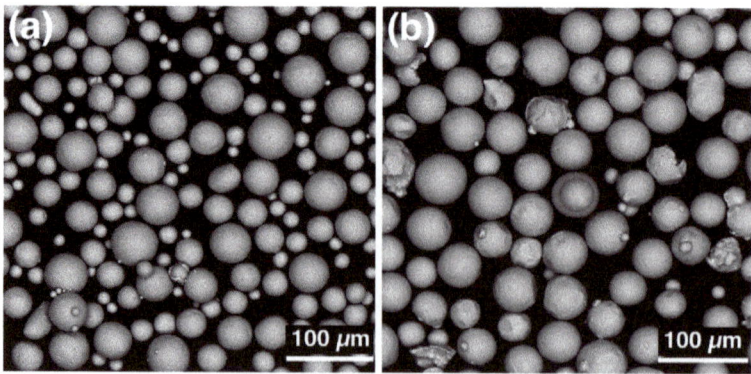

Figure 1. Scanning electron microscope (SEM)-images of (**a**) Ti and (**b**) Ti64 powders.

The samples were fabricated using the SLM Solutions 280HL machine (Lübeck, Germany) in an argon atmosphere (99.99% purity) on a Ti64 built substrate. For microstructural characterization and microhardness evaluation, the samples of 20 mm height and

15 × 15 mm² section were built. Initially, one of the feedstock materials was used during the L-PBF process to fabricate the first half of the sample (10 mm). After that, the powder in the machine was changed to the second material and the second half of the sample was manufactured. The samples for mechanical tests were fabricated using a similar technique by changing the powder in the machine after building part of the sample. The same L-PBF process parameters were used for Ti and Ti64 alloys that were chosen based on the previous studies [28,29]. The following process parameters were used: scanning speed—805 mm/s, laser power—275 W, hatch distance—120 µm, layer thickness—50 µm. The laser beam size was approximately 80 µm.

Heat treatment of the samples was carried out using a vacuum furnace (Carbolite Gero GmbH & Co. KG, Neuhausen, Germany) at 10^{-3}–10^{-4} mbar at 950 °C for 2 h, followed by furnace cooling. The regime was chosen based on AMS-H-81200A specification for the Ti64 alloy. The same temperature was used for hot isostatic pressing (HIP) of the graded samples, while the pressure was 100 MPa.

The microstructure was studied using a Leica DMI 5000 (Leica Microsystems, Wetzlar, Germany) optical microscope. To study the chemical composition, a Mira 3 (TESCAN, Brno, Czech Republic) scanning electron microscope with an energy dispersive X-ray (EDX) spectroscopy module was used.

Ti/Ti64 samples were scanned on a v | tome | x m300 X-ray computer tomography (CT). The system was equipped with an X-ray source with a maximum voltage of 300 kV. The obtained data were processed and visualized using an extended software package AVIZO for three-dimensional analysis and voxels visualization. Segmentation was performed using global and local gray thresholds.

The hardness of the samples was measured using Zwick/Roell ZHU 250 tester (Zwick GmbH, Ulm, Germany) with a Vickers intender along the material transition area.

Tensile tests were carried out at room temperature using Zwick/Roell Z050 (Zwick GmbH, Ulm, Germany) testing machine. Figure 2 schematically shows the samples used for tensile tests of the graded materials. The gauge length was 45 mm and the width was 20 mm, while the thickness of the specimens was 3 mm. Three samples per point were used for the tensile tests.

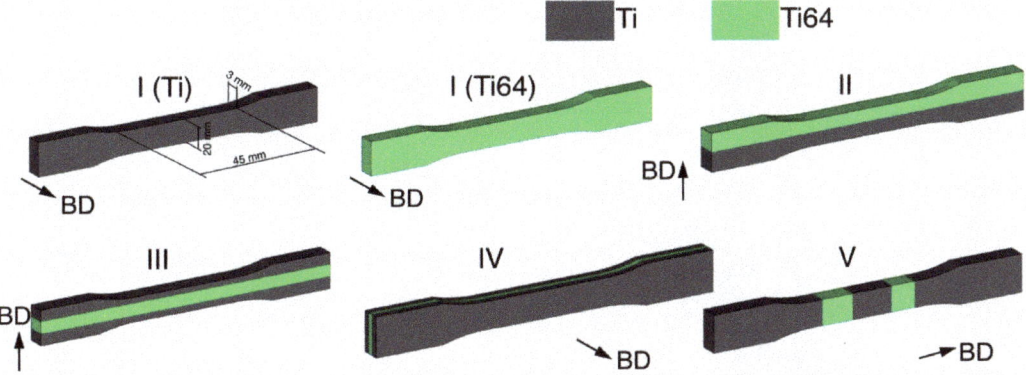

Figure 2. Schematic representation of tensile specimens' configuration.

Specimens consisting entirely of Ti of Ti64 materials were labeled as I (Ti) and I (Ti64), respectively. The remaining specimens consisting of two materials are shown in Figure 2. Type II specimens were split in half and consisted of 50% Ti and 50% Ti64 materials. Type III and IV had an insert from Ti64 alloy located at the center of the specimen with different orientations of the insert with 50%/50% volume fraction of the alloys. The IV type specimen had two inserts from Ti64 alloy as shown in Figure 2.

3. Results and Discussion

A sample consisting of Ti and Ti64 materials fabricated by L-PBF is shown in Figure 3a. There are no visible differences between the zones of the sample externally. They have the same color and surface roughness. A small line along the alloy interface, caused by the thermal expansion of the lower zone during fabrication, is noticeable. An image of the transition zone section of the sample in the initial state obtained by computed tomography (Figure 3c) shows internal defects in the form of pores, the average size of which is about 50 μm. It is also possible to see the transition from one material to another using the computed tomography, as the Ti64 material has a lighter shade compared to pure Ti due to the difference in density. The residual pores in the material after the HIP were not detected by CT, but the transition zone from one material to the other can also be seen (Figure 3d).

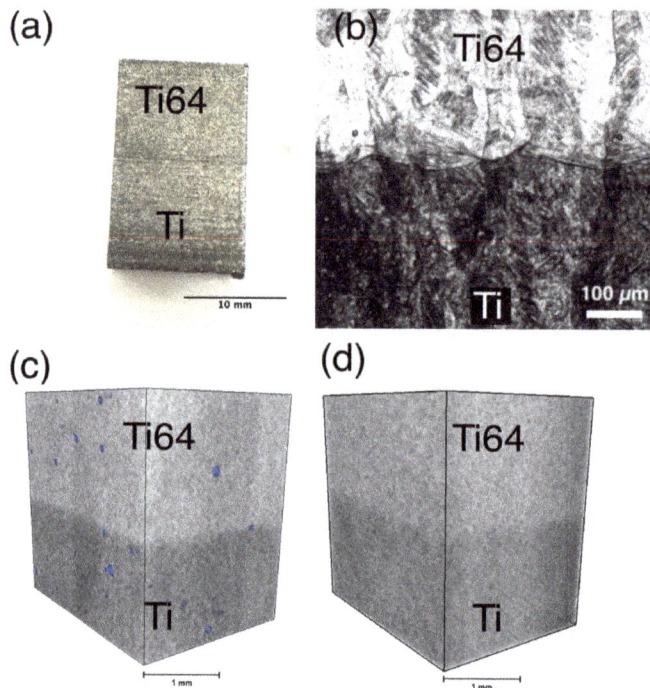

Figure 3. (a) A photograph of the Ti/Ti64 sample, (b) microstructure of the transition zone, (c) as-built, and (d) HIPed specimens' volume obtained by CT-reconstruction.

The microstructure of the transition zone between the two materials is shown in Figure 3b. No visible defects in the form of a lack of fusion or cracks were found. After etching, a distinct transition zone can be seen between Ti and Ti64 zones. It can be seen that the transition zone has a thickness of about 50–100 μm, which corresponds to the 1–2 layer thickness used during the L-PBF process.

The microstructure of the Ti zone consists of fine martensitic α' needles, while the Ti64 zone exhibits a needle-like martensitic α' phase within the columnar primary β grains. The high solidification rates typical for the L-PBF process resulted in a metastable microstructure in the case of both alloys. Titanium undergoes an $\alpha \rightarrow \beta$ phase transformation above 890 °C, and this allotropic phase transformation affects the microstructure and texture of the material. The Ti64 alloy undergoes an $\beta \leftrightarrow \alpha + \beta$ phase transformation at about 1000 °C [30]. However, the L-PBF process leads to metastable martensitic microstructure due to the high cooling rates up to 10^5 K/s [31]. The L-PBF process, accompanied by rapid solidification, leads to the formation of a martensitic microstructure with elongated grains of the primary β-phase filled with the finely dispersed lamellar α-phase. The partial remelting of the previous layers provides the epitaxial growth of such grains.

The results of a study of the chemical composition of the transition zone (Figure 4) showed that the Al and V content increases smoothly from the zone of Ti to the zone of Ti64. According to the measurement results, the width of the transition zone can be estimated at approximately 200 μm.

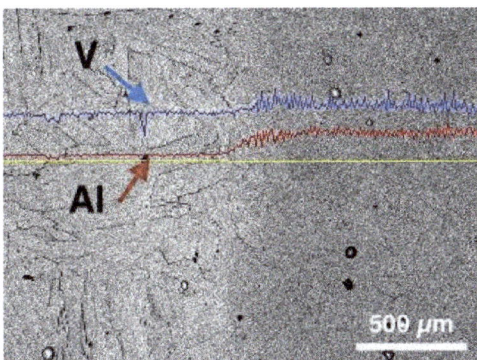

Figure 4. EDX results showing the change of V and Al composition distribution along the transition zone from Ti (**left**) to Ti64 (**right**) on the sample.

Depending on the heat treatment temperature and cooling rate, the titanium microstructure may have different morphology: equiaxed α-Ti grains inside the primary β-grains, a Widmanstett structure, and a lamellar or needle morphology of the α-phase [32].

The heat treatment of the Ti/Ti64 sample at 950 °C resulted in the transition of the martensitic α′-phase to α + β phase in the case of Ti64 zone and α-Ti phase grains in the case of Ti zone (Figure 5).

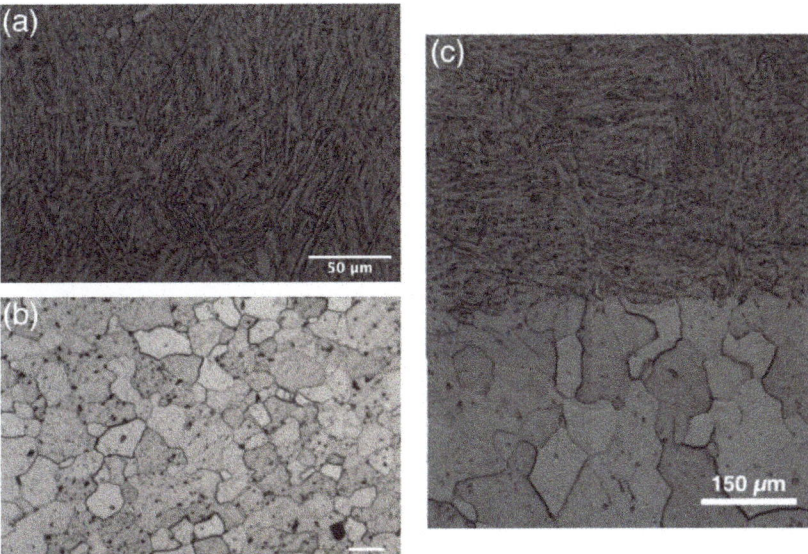

Figure 5. Microstructures of Ti/Ti64 sample after heat treatment: (**a**) Ti64 zone, (**b**) Ti zone, (**c**) the transition zone.

After heat treatment, the Ti zone consists of equiaxed α-Ti grains, which indicate the recrystallization processes [33]. The recrystallization process led to the disappearance of the preferential orientation of the grains. They had an equiaxial shape and size from 80 to 150 µm.

After HIP, the microstructure of the samples underwent changes similar to those after the heat treatment, but there were differences in the morphology. The microstructure of the Ti64 alloy zone (Figure 6a) also consists of α + β phases with lamellar morphology, formed as a result of martensitic α'-phase decomposition to α + β. The formation of the lamellar α-phase with a larger lamellar size and β-phase grains occurs in the Ti64 alloy, in comparison to heat-treated conditions. The increase in the α-phase lamellar size occurs both within and along the grain boundaries, which may lead to an increase in ductility as deformation is mainly found along the grain boundaries.

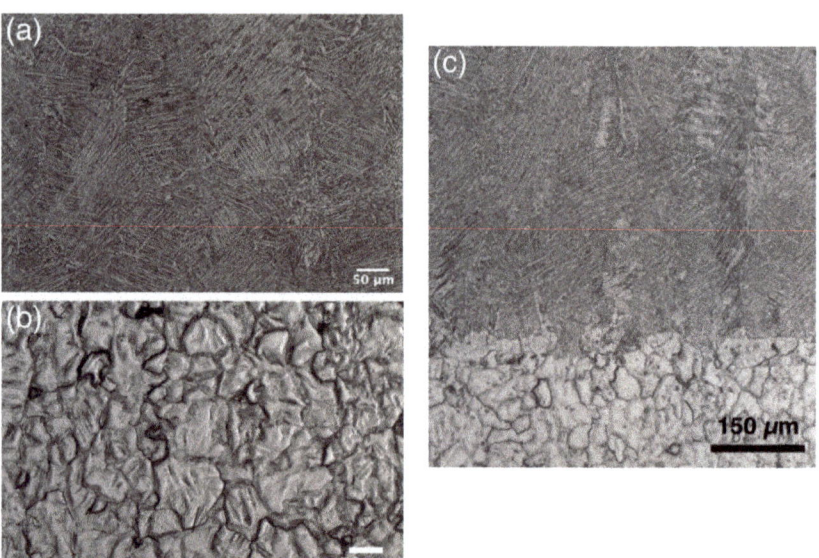

Figure 6. Microstructures of Ti/Ti64 sample after HIP: (**a**) Ti64 zone, (**b**) Ti zone, and (**c**) the transition zone.

The Ti zone (Figure 6b) after HIP has a microstructure of equiaxed α-Ti grains with larger sizes compared to heat-treated ones, which may be caused by the differences in the cooling rate at a different post-processing [29].

Figure 7 shows hardness distribution for the Ti/Ti64 samples along the material transition area of as-built, heat-treated, and HIPed samples.

The hardness of the Ti64 zone is higher compared to the Ti zone for all tested conditions. The as-built condition exhibited the highest hardness values for both the Ti and Ti64 zone due to metastable microstructures formed during the L-PBF process. After heat treatment and hot isostatic pressing, the hardness values decreased for both the Ti and Ti64 zones due to the stress relieving and martensite phase decomposition. Due to different cooling conditions in the vacuum furnace and with HIP, there were differences between the values of hardness. After HIP, the microstructure was slightly coarser in terms of grain and lamellar size compared to the heat-treated samples, which resulted in lower hardness values along all the zones for the HIPed sample.

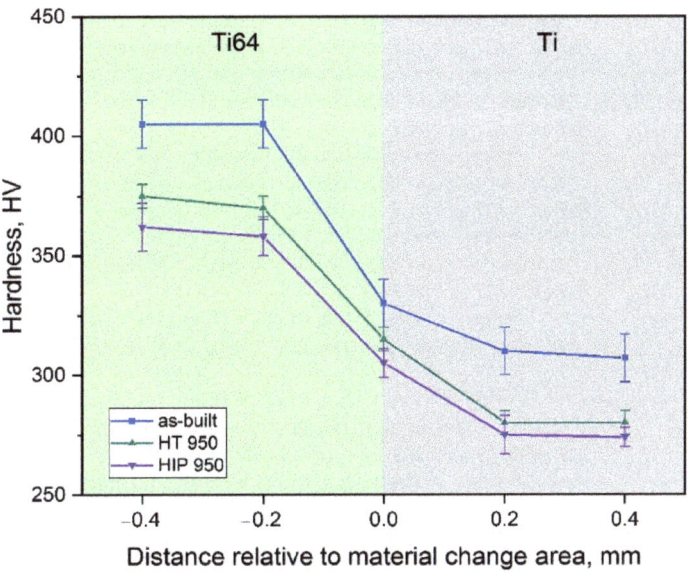

Figure 7. Hardness distribution for Ti/Ti64 samples along the material change area for as-built, heat treated, and HIPed conditions.

Tensile tests of pure Ti, Ti64, and graded Ti/Ti64 material have been made for samples fabricated by L-PBF and subsequent hot isostatic pressing. The results are summarized in Table 1.

Table 1. The results of tensile tests of Ti, Ti64, and graded Ti/Ti64 samples produced by L-PBF with the subsequent HIP.

Specimen Type	Tensile Strength, MPa	Yield Strength, MPa	Elongation at Break, %
I (Ti)	700 ± 12	596 ± 8	16 ± 4
I (Ti64)	998 ± 21	821 ± 11	10 ± 3
II	728 ± 29	693 ± 25	3 ± 1
III	760 ± 14	726 ± 11	5 ± 2
IV	839 ± 10	754 ± 8	7 ± 3
V	703 ± 36	667 ± 32	2 ± 1

The values of yield and tensile strengths of all the graded samples were between the values for pure Ti and Ti64. The samples of the V type had the lowest strength values; this type had 4 Ti/Ti64 material change interfaces which transversely directed the forces applied during the tensile test. The type II samples turned out to be more strengthened; this type had one interface of Ti/Ti64 material changing, directed along the axis of the tensile during the test. Even the higher strength values were demonstrated by type III samples. This type was characterized by the 2 Ti/Ti64 material change interfaces along the axis of the tensile and the total area of the interfaces were twice as large as those of type II. The highest strength characteristics between the investigated graded materials were found in the type IV samples. The geometry of this sample type had 2 Ti/Ti64 material change interfaces along the axis of the tensile with the largest contact area between the different materials.

The graded samples did not exhibit high elongation values. The changes in the elongation values for different types of samples have a similar tendency to the strength values. The fracture of the type V samples occurred at the material interface. This was potentially due to a partial oxidation of the metal surface, or the cooling of the specimens during a material change. Other sample types had interfaces along the axis of the tensile, as well as a low elongation. Therefore, another possible reason for the low elongation may be the Ti64 zones having a higher yield strength which could limit the elongation of the Ti zones and lead to the formation and failure of stress concentrators with relatively low values of elongation.

It should be noted that the homogeneous specimens, as well as the type V specimens, were fabricated with a build direction along the tensile load, and the other specimen types were fabricated with a build direction perpendicular to the tensile load. As shown in the previous research [34], the strength of the horizontally fabricated samples was higher compared to the vertically built samples, which could contribute to the lower properties of the type V specimens.

The presented results of the investigation of the graded Ti/Ti64 samples allowed tailoring properties for possible applications of multimaterial parts.

4. Conclusions

The investigation of samples with the graded chemical composition, Ti/Ti64, was presented in this work. The study of the transition zone structure showed that these samples had a distinct character and did not undergo significant changes during heat treatment and hot isostatic pressing.

The study of tensile mechanical properties showed that, when the zones are located along the sample, the ratio of cross-sections has a greater influence on the mechanical properties than their shape and location. When the zones are arranged transversely to the specimen, a failure occurs at the interface and the relative elongation is extremely low. Future investigations in multimaterial 3D printing must pay attention to the possibility of creating change interfaces with the smooth changing of chemical compositions and an increasing transient zone.

Author Contributions: Conceptualization, E.B., I.P. and V.S.; Data curation, K.S.; Investigation, E.B., I.P. and K.S.; Project administration, E.B. and V.S.; Resources, A.P.; Visualization, E.B.; Writing—original draft, E.B. and V.S.; Writing—review and editing, I.P., E.B. and V.S. All authors have read and agreed to the published version of the manuscript.

Funding: The research is partially funded by the Ministry of Science and Higher Education of the Russian Federation as part of World-class Research Center program: Advanced Digital Technologies (contract No. 075-15-2020-934 dated 17 November 2020).

Institutional Review Board Statement: Not applicable.

Informed Consent Statement: Not applicable.

Data Availability Statement: The data presented in this study are available on request from the corresponding author.

Conflicts of Interest: The authors declare no conflict of interest.

References

1. Orme, M.E.; Gschweitl, M.; Ferrari, M.; Madera, I.; Mouriaux, F. Designing for Additive Manufacturing: Lightweighting Through Topology Optimization Enables Lunar Spacecraft. *J. Mech. Des.* **2017**, *139*, 100905. [CrossRef]
2. Gao, W.; Zhang, Y.; Ramanujan, D.; Ramani, K.; Chen, Y.; Williams, C.B.; Wang, C.C.L.; Shin, Y.C.; Zhang, S.; Zavattieri, P.D. The status, challenges, and future of additive manufacturing in engineering. *Comput. Des.* **2015**, *69*, 65–89. [CrossRef]
3. Uriondo, A.; Esperon-Miguez, M.; Perinpanayagam, S. The present and future of additive manufacturing in the aerospace sector: A review of important aspects. *Proc. Inst. Mech. Eng. Part G J. Aerosp. Eng.* **2015**, *229*, 2132–2147. [CrossRef]
4. Zhang, L.-C.; Attar, H.; Calin, M.; Eckert, J. Review on manufacture by selective laser melting and properties of titanium based materials for biomedical applications. *Mater. Technol.* **2015**, *7857*, 1–11. [CrossRef]
5. Popovich, A.A.; Sufiiarov, V.S.; Borisov, E.V.; Polozov, I.A.; Masaylo, D.V. Design and manufacturing of tailored microstructure with selective laser melting. *Mater. Phys. Mech.* **2018**, *38*, 1–10.
6. Singh, R.; Kumar, R.; Farina, I.; Colangelo, F.; Feo, L.; Fraternali, F. Multi-Material Additive Manufacturing of Sustainable Innovative Materials and Structures. *Polymers* **2019**, *11*, 62. [CrossRef]

7. Han, D.; Lee, H. Recent advances in multi-material additive manufacturing: Methods and applications. *Curr. Opin. Chem. Eng.* **2020**, *28*, 158–166. [CrossRef]
8. Bandyopadhyay, A.; Heer, B. Additive manufacturing of multi-material structures. *Mater. Sci. Eng. R Reports* **2018**, *129*, 1–16. [CrossRef]
9. Jing, S.; Zhang, H.; Zhou, J.; Song, G. Optimum weight design of functionally graded material gears. *Chin. J. Mech. Eng.* **2015**, *28*, 1186–1193. [CrossRef]
10. Singh, A.K. A novel technique for in-situ manufacturing of functionally graded materials based polymer composite spur gears. *Polym. Compos.* **2019**, *40*, 523–535. [CrossRef]
11. Arabnejad, S.; Johnston, R.B.; Pura, J.A.; Singh, B.; Tanzer, M.; Pasini, D. High-Strength Porous Biomaterials for Bone Replacement: A strategy to assess the interplay between cell morphology, mechanical properties, bone ingrowth and manufacturing constraints. *ACTA Biomater.* **2015**, *30*, 345–356. [CrossRef] [PubMed]
12. Walker, J.; Middendorf, J.R.; Lesko, C.C.C.; Gockel, J. Multi-material laser powder bed fusion additive manufacturing in 3-dimensions. *Manuf. Lett.* **2021**, in press. [CrossRef]
13. Wei, C.; Li, L.; Zhang, X.; Chueh, Y.-H. 3D printing of multiple metallic materials via modified selective laser melting. *CIRP Ann.* **2018**, *67*, 245–248. [CrossRef]
14. Sing, S.L.; Huang, S.; Goh, G.D.; Goh, G.L.; Tey, C.F.; Tan, J.H.K.; Yeong, W.Y. Emerging metallic systems for additive manufacturing: In-situ alloying and multi-metal processing in laser powder bed fusion. *Prog. Mater. Sci.* **2021**, *119*, 100795. [CrossRef]
15. Wei, C.; Gu, H.; Sun, Z.; Cheng, D.; Chueh, Y.-H.; Zhang, X.; Huang, Y.; Li, L. Ultrasonic material dispensing-based selective laser melting for 3D printing of metallic components and the effect of powder compression. *Addit. Manuf.* **2019**, *29*, 100818. [CrossRef]
16. Wei, C.; Gu, H.; Zhang, X.; Chueh, Y.; Li, L. Hybrid ultrasonic and mini-motor vibration-induced irregularly shaped powder delivery for multiple materials additive manufacturing. *Addit. Manuf.* **2020**, *33*, 101138. [CrossRef]
17. Sorkin, A.; Tan, J.L.; Wong, C.H. Multi-material modelling for selective laser melting. *Procedia Eng.* **2017**, *216*, 51–57. [CrossRef]
18. Zhang, M.; Yang, Y.; Wang, D.; Song, C.; Chen, J. Microstructure and mechanical properties of CuSn/18Ni300 bimetallic porous structures manufactured by selective laser melting. *Mater. Des.* **2019**, *165*, 107583. [CrossRef]
19. Bartolomeu, F.; Costa, M.M.; Alves, N.; Miranda, G.; Silva, F.S. Additive manufacturing of NiTi-Ti6Al4V multi-material cellular structures targeting orthopedic implants. *Opt. Lasers Eng.* **2020**, *134*, 106208. [CrossRef]
20. Scaramuccia, M.G.; Demir, A.G.; Caprio, L.; Tassa, O.; Previtali, B. Development of processing strategies for multigraded selective laser melting of Ti6Al4V and IN718. *Powder Technol.* **2020**, *367*, 376–389. [CrossRef]
21. Khaimovich, A.; Erisov, Y.; Smelov, V.; Agapovichev, A.; Petrov, I.; Razhivin, V.; Bobrovskij, I.; Kokareva, V.; Kuzin, A. Interface Quality Indices of Al–10Si–Mg Aluminum Alloy and Cr18–Ni10–Ti Stainless-Steel Bimetal Fabricated via Selective Laser Melting. *Metals* **2021**, *11*, 172. [CrossRef]
22. Chen, J.; Yang, Y.; Song, C.; Zhang, M.; Wu, S.; Wang, D. Interfacial microstructure and mechanical properties of 316L /CuSn10 multi-material bimetallic structure fabricated by selective laser melting. *Mater. Sci. Eng. A* **2019**, *752*, 75–85. [CrossRef]
23. Rankouhi, B.; Jahani, S.; Pfefferkorn, F.E.; Thoma, D.J. Compositional grading of a 316L-Cu multi-material part using machine learning for the determination of selective laser melting process parameters. *Addit. Manuf.* **2021**, *38*, 101836.
24. Neirinck, B.; Li, X.; Hick, M. Powder Deposition Systems Used in Powder Bed-Based Multimetal Additive Manufacturing. *Accounts Mater. Res.* **2021**, *2*, 387–393. [CrossRef]
25. Schneck, M.; Horn, M.; Schmitt, M.; Seidel, C.; Schlick, G.; Reinhart, G. Review on additive hybrid- and multi-material-manufacturing of metals by powder bed fusion: State of technology and development potential. *Prog. Addit. Manuf.* **2021**, *2021*, 1–14.
26. Heat Exchenager—Aerosint. Available online: https://aerosint.com/heat-exchanger/ (accessed on 14 October 2021).
27. Saji, V.S.; Jeong, Y.H.; Choe, H.C. A Comparative Study on Corrosion Behavior of Ti-35Nb-5Ta-7Zr Ti-6Al-4V and CP-Ti in 0.9 wt% NaC. *Corros. Sci. Technol.* **2009**, *8*, 139–142.
28. Sufiiarov, V.S.; Popovich, A.A.; Borisov, E.V.; Polozov, I.A. Selective laser melting of titanium alloy and manufacturing of gas-turbine engine part blanks. *Tsvetnye Met.* **2015**, *8*, 76–80. [CrossRef]
29. Popovich, A.; Sufiiarov, V.; Borisov, E.; Polozov, I. Microstructure and Mechanical Properties of Ti-6Al-4V Manufactured by SLM. *Key Eng. Mater.* **2015**, *651–653*, 677–682. [CrossRef]
30. Tarín, P.; Gualo, A.; Simón, A.G.; Piris, N.M.; Badía, J.M. Study of Alpha-Beta Transformation in Ti-6Al-4V-ELI. Mechanical and Microstructural Characteristics. *Mater. Sci. Forum* **2010**, *638–642*, 712–717. [CrossRef]
31. Pauly, S.; Wang, P.; Kühn, U.; Kosiba, K. Experimental determination of cooling rates in selectively laser-melted eutectic Al-33Cu. *Addit. Manuf.* **2018**, *22*, 753–757. [CrossRef]
32. Ibrahim, K.M.; Mhaede, M.; Wagner, L. Effect of Annealing Temperature on Microstructure and Mechanical Properties of Hot Swaged cp-Ti Produced by Investment Casting. *J. Mater. Eng. Perform.* **2012**, *21*, 114–118. [CrossRef]
33. Li, C.-L.; Won, J.W.; Choi, S.-W.; Choe, J.-H.; Lee, S.; Park, C.H.; Yeom, J.-T.; Hong, J.K. Simultaneous achievement of equiaxed grain structure and weak texture in pure titanium via selective laser melting and subsequent heat treatment. *J. Alloys Compd.* **2019**, *803*, 407–412. [CrossRef]
34. Rafi, H.K.; Karthik, N.V.; Gong, H.; Starr, T.L.; Stucker, B.E. Microstructures and Mechanical Properties of Ti6Al4V Parts Fabricated by Selective Laser Melting and Electron Beam Melting. *J. Mater. Eng. Perform.* **2013**, *22*, 3872–3883. [CrossRef]

Article

Structure and Properties of Barium Titanate Lead-Free Piezoceramic Manufactured by Binder Jetting Process

Vadim Sufiiarov *, Artem Kantyukov, Anatoliy Popovich and Anton Sotov

Institute of Mechanical Engineering, Materials, and Transport, Peter the Great St. Petersburg Polytechnic University, 195251 Saint Petersburg, Russia; kantyukov.artem@mail.ru (A.K.); director@immet.spbstu.ru (A.P.); SotovAnton@yandex.ru (A.S.)
* Correspondence: vadim.spbstu@yandex.ru

Abstract: This article presents the results of manufacturing samples from barium titanate (BaTiO$_3$) lead-free piezoceramics by using the binder jetting additive manufacturing process. An investigation of the manufacturing process steps for two initial powders with different particle size distributions was carried. The influence of the sintering and the particle size distribution of the starting materials on grain size and functional properties was evaluated. Samples from fine unimodal powder compared to coarse multimodal one have 3–4% higher relative density values, as well as a piezoelectric coefficient of 1.55 times higher values (d$_{33}$ = 183 pC/N and 118 pC/N correspondingly). The influence of binder saturation on sintering modes was demonstrated. Binder jetting with 100% saturation for both powders enables printing samples without delamination and cracking. Sintering at 1400 °C with a dwell time of 6 h forms the highest density samples. The microstructure of sintered samples was characterized with scanning electron microscopy. The possibility of manufacturing parts from functional ceramics using additive manufacturing was demonstrated.

Keywords: additive manufacturing; binder jetting; lead-free piezoceramic; barium titanate; sintering; piezoelectric properties

1. Introduction

Functional ceramics are a class of materials that exhibit special properties in addition to those already inherent in ceramics, such as chemical and thermal stability. Functional ceramics typically exhibit one or more unique properties: biological, electrical, magnetic, or chemical. [1]. Due to this, they are used in engine production, aviation, and space industries [2]. The most promising types of functional ceramics include piezoceramics [3]. Piezoelectric ceramics are used to make sensor devices, energy harvesters, and actuators [4–6]. Piezoelectric materials are of particular interest as pressure and temperature sensors in high-frequency environments [7,8]. Piezoelectric ceramics have generated particular interest in the power industry because they can withstand the harsh environmental conditions present in energy conversion systems [9]. However, despite their advantages as sensitive devices, piezoceramics also have the same internal disadvantages that are observed in most ceramic materials: they are difficult to process [10], and their fragility causes low fracture resistance [11]. Therefore, the manufacture of non-standard complex geometries from ceramic materials can be practically impossible using conventional manufacturing methods. The proposed method for circumventing this problem is the manufacture of complex ceramic parts using additive manufacturing (AM) [12]. AM has such advantages as the absence of expensive tools, easy scalability of the process, the ability to implement parts of complex shapes, a high degree of material utilization and minimum production time [13]. One of the most relevant materials for ceramic additive manufacturing is a piezoelectric material since it generates an electric charge when deformed or, conversely, deforms when an electric potential is applied. The use of AM for the manufacture of piezoelectric materials will expand the scope of their application, expanding the

Citation: Sufiiarov, V.; Kantyukov, A.; Popovich, A.; Sotov, A. Structure and Properties of Barium Titanate Lead-Free Piezoceramic Manufactured by Binder Jetting Process. *Materials* **2021**, *14*, 4419. https://doi.org/10.3390/ma14164419

Academic Editor: Haim Abramovich

Received: 18 June 2021
Accepted: 4 August 2021
Published: 6 August 2021

Publisher's Note: MDPI stays neutral with regard to jurisdictional claims in published maps and institutional affiliations.

Copyright: © 2021 by the authors. Licensee MDPI, Basel, Switzerland. This article is an open access article distributed under the terms and conditions of the Creative Commons Attribution (CC BY) license (https://creativecommons.org/licenses/by/4.0/).

possibilities of forming multilayer, as well as complex geometries of structures. Therefore, the possibility of integrating piezoelectric materials during the manufacturing process itself would lead to the creation of multifunctional structures within a single processing process. With great freedom in the achievable geometries of piezoelectric elements, the prospect opens for a significant improvement in the performance of many devices based on piezoelectric and ferroelectric properties [14,15].

Some previous investigations of additive manufacturing lead-based piezoceramics have been made using direct writing/FDM [16,17], stereolithography-based processes [18,19], and ink-jetting [20]. Due to the toxicity of lead compounds, the development of new piezomaterials and technologies is moving towards lead-free piezomaterials. Barium titanate ($BaTiO_3$) is one of the most widely used lead-free piezoceramic materials, which became widespread due to its high dielectric and piezoelectric properties [21]. The most promising methods of 3D printing $BaTiO_3$ are direct writing (DW) [22–25], vat photopolymerization (VP) [26–30], and binder jetting (BJ) [31–34]. Samples printed using DW have the best piezoelectric coefficient values (d_{33} = 200 pC/N [22]) and a density (6.01 g/cm^3 [23]) close to the theoretical density limit of $BaTiO_3$ (6.02 g/cm^3). However, the quality of the surface printed layer is rather rough, which may be a limitation for using this technology. It is also worth noting that difficulties arise when using printing nozzles with a diameter less than 500 microns—the nozzle can clog with ceramic powder particles, and this reduces the accuracy of printing parts [23]. Samples printed using VP also have high piezoelectric coefficient values (d_{33} = 165 pC/N [27] and high material density (5.64 g/cm^3 [26]). However, there are next limitations when using this 3D printing technology [29]: (i) using 3D printer with a layer spreading system for viscous slurry with a high solid loading of ceramics in the photopolymer resin; (ii) the high refractive index of UV light for $BaTiO_3$ that limits the curing depth; (iii) a long time of debinding process that directly affects the final result of the subsequent sintering. Samples printed using BJ have low piezoelectric coefficient values (d_{33} = 74.1 pC/N [31], d_{33} = 112 pC/N [32] and have a low density (3.93 g/cm^3 [31], 2.21 g/cm^3 [32]). However, this technology ensures the high quality of printed parts and excludes difficulties of the debinding process.

The BJ additive process is a method where a nozzle print-head jets a liquid binder on a powder layer in places that correspond to the cross-section of the computer model of a part. The result of BJ printing is a green model with low mechanical properties and high porosity. The green model needs further curing, debinding, and sintering. As a result, the characteristics of parts made of the polymer [35], metal [36–38] and ceramics [39] printed on a 3D printer largely depend on manufacturing and postprocessing parameters. Consequently, the behavior of functional ceramics made with additive technologies must be further studied to expand the capabilities of this new technique.

In this paper, $BaTiO_3$ lead-free piezoceramic was used to study the additive manufacturing of piezoelements by using the BJ process. The influence of the manufacturing process on the properties of the material was characterized and discussed, and the dielectric and piezoelectric properties of the manufactured samples were measured.

2. Materials and Methods

2.1. Materials

Two types of $BaTiO_3$ powder were used for printing by BJ: (i) micron powder with multimodal particle size distribution (PSD) (C-$BaTiO_3$, ZAO NPF Luminofor, Stavropol, Russia) D_{10}—0.1 μm, D_{50}—3.4 μm, D_{90}—25.4 μm, and (ii) submicron powder with unimodal PSD (F-$BaTiO_3$, Acros Organics, Geel, Belgium) D_{10}—0.6 μm, D_{50}—1.1 μm, D_{90}—2.1 μm. Figures 1 and 2 shows images of as-received powders. The C-$BaTiO_3$ powder has particle sizes of about 1 μm, forms agglomerates up to 25 μm (Figure 1). The F-$BaTiO_3$, powder has particle sizes about 1 μm (Figure 2). Figure 3 shows the particle size distributions which demonstrates that the C-$BaTiO_3$ powder consists of agglomerates and reveals three peaks and contains small particles. The unimodal powder is much more homogeneous, while in the case of the multimodal powder one can observe the agglomeration of small particles into larger clusters, which correspond to the third peak in the PSD with a medium size at about 20 microns.

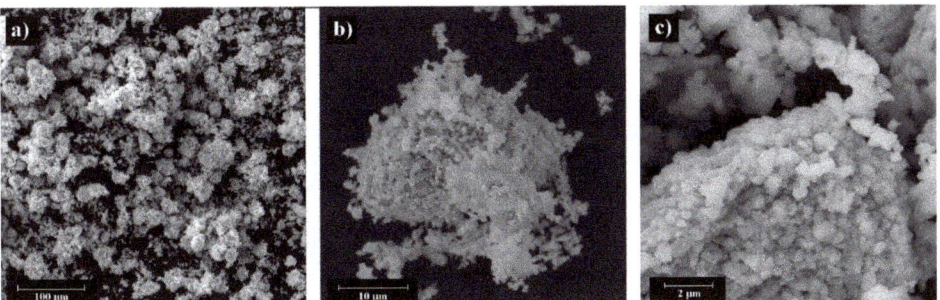

Figure 1. SEM images of C-BaTiO$_3$ powder at different magnifications: general view (**a**), agglomerate (**b**), powder particles (**c**).

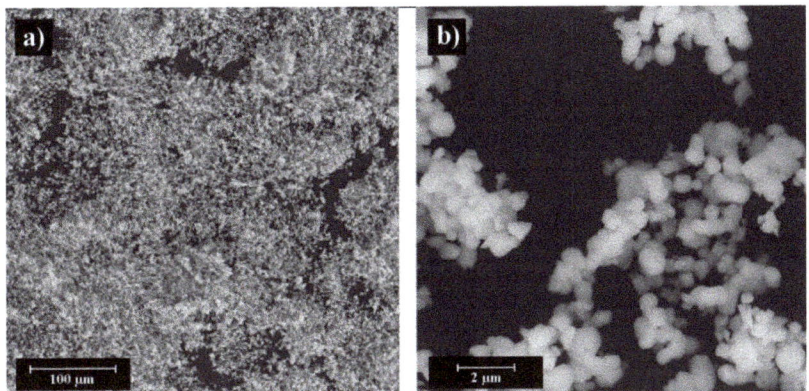

Figure 2. SEM images of F-BaTiO$_3$ powder at different magnifications: general view (**a**), powder particles (**b**).

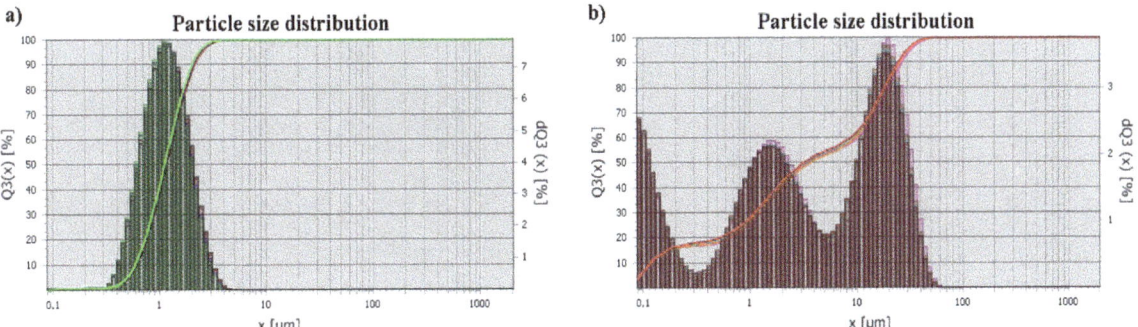

Figure 3. Particle size distribution of F-BaTiO$_3$ (**a**) and C-BaTiO$_3$ (**b**).

2.2. Fabrication

For the experiment green models of cubic shape with dimensions of $10 \times 10 \times 10$ mm^3 were printed to study subsequent debinding and sintering processes. Also, the two types of cylindrical green models (diameter 15 mm and 10 mm, height 10 mm and 1 mm respectively) were printed to investigate the electromechanical properties.

The piezoceramic samples were manufactured on the ExOne Innovent system (The ExOne Company, North Huntingdon, PA, USA). This system relates to the BJ additive manufacturing process. The original ExOne BS004 solvent binder and CL001 cleaner were used for the printing of the functional ceramic components.

The BJ process can be divided into several stages, a schematic image of which is shown in Figure 4:

1. A thin layer of powder material is formed on the platform using a roller;
2. A liquid binder is selectively sprayed to the powder layer using a print head, in accordance with the cross-section of the computer model;
3. Then the platform is lowered to a given thickness of one layer;
4. The powder layer is dried and heated using an infrared heater;
5. From the hopper, using an oscillator, the powder is fed to the surface of the platform and a new layer of powder is applied;
6. Then the layer is leveled using a rotating roller;
7. Processes 1–6 are repeated until a full-size green model is made.

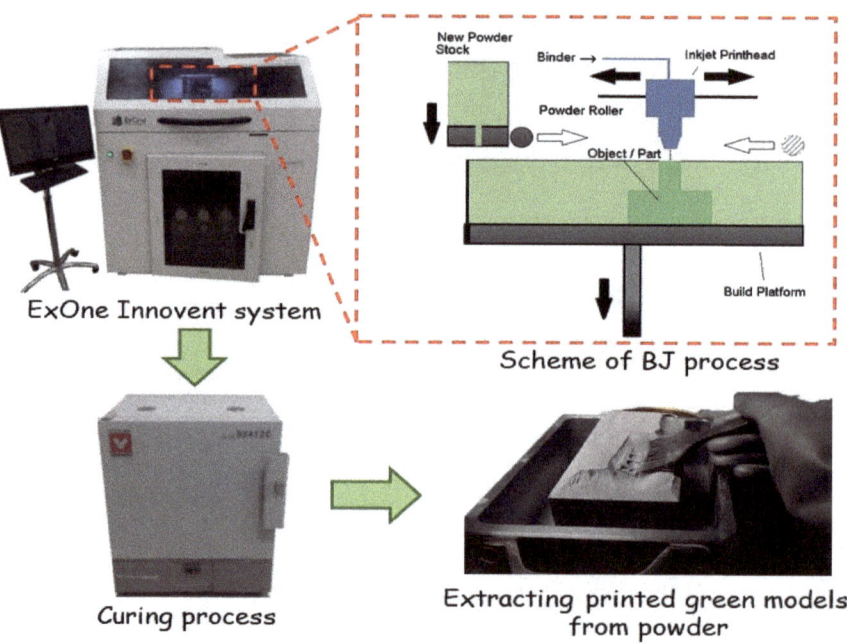

Figure 4. Process flow for BJ additive manufacturing.

Printed parts are considered "green" and are not suitable for end-use. Thus, these green models require further post-processing, such as sintering or infiltration, to achieve the desired mechanical and functional properties.

After the 3D printing, the platform (together with the green models in the powder surround) is placed in the thermal furnace (Yamato DX412C, Yamato Scientific, Santa Clara, CA, USA) at 180 °C for 3 h for curing.

After curing, the green models have sufficient strength to remove excess and loose powder. For green models of a simple shape, removal was done with a brush; for complex shapes removal was done using compressed air.

2.3. Thermal Post-Treatment

Before thermal post-treatment, the green models were placed in alumina crucibles with lids. The debinding process was performed in a muffle furnace (KJ-1700X, Zhengzhou Kejia Furnace Co., Ltd., Zhengzhou, China) at a temperature of 650 °C with a dwell time of 60 min. After debinding, the samples were sintered in a muffle furnace at 1300, 1350, and 1400 °C with a dwell times of 2, 4, and 6 h under an air atmosphere. The thermal post-treatment profiles are illustrated in Figure 5.

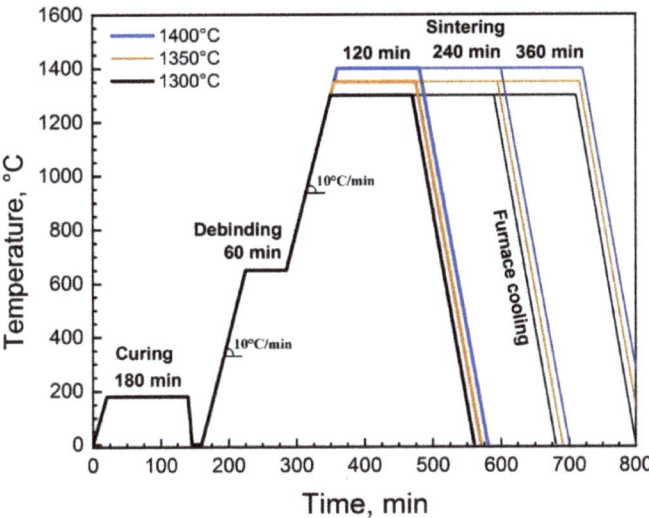

Figure 5. The thermal post-treatment profile.

2.4. Characterization

The particle size distribution of the powders was determined by laser diffraction Analysette 22 NanoTec plus (Fritsch, Idar-Oberstein, Germany) with a total measurement range of 0.01–2000 µm.

TGA analysis of BS004 solvent binder was performed using a thermogravimetric analyzer (Q5000, TA Instruments, New Castle, DE, USA). The heating was carried out in an airflow of 30 mL/min in the temperature range 30–700 °C at a rate of 10 °C/min. The binder was placed in a platinum crucible, after which it was heated from room temperature up to 700 °C in air.

The structure of the samples after sintering was studied using a Leica DMI5000 optical microscope (Leica, Wetzlar, Germany) and a Tescan Mira3 LMU scanning electron microscope (SEM) operating at magnifications from 4× to $10^6 \times$ with an accelerating voltage from 200 V to 30 kV. The chemical composition was measured using an energy-dispersion accessory into the SEM.

Optical and SEM-images of sintered samples from C-BaTiO$_3$ and F-BaTiO$_3$ were examined using the ImageJ Software. v.1.52a (Bethesda, MD, USA) The grain sizes were analyzed for various temperature and time sintering conditions.

The density of the sintered samples was measured by the Archimedes method; the calculation of relative density was made in accordance with the theoretical density of BaTiO$_3$ (6.02 g/cm^3).

The phase composition was analyzed using a Bruker D8 Advance X-ray (Bruker corp., Billerica, MA, USA) diffractometer (XRD) using CuKa radiation (l = 1.5418 Å) without monochromator.

All samples for the electrical performance test were coated with silver electrodes (paste PP-17, Delta, Zelenograd, Russia) at 700 °C for 30 min. The samples were poled in air, at Tc + 20 °C (Tc-Curie temperature 120–130 °C for $BaTiO_3$). Then, an electric field of 0.6 kV/mm for 30 min was applied to samples, followed by cooling to room temperature. Dielectric constant ε, the loss tangent $tg\delta$, electromechanical coupling coefficient k_p, and piezoelectric coefficient d_{33} were measured and calculated. Dielectric properties were measured on cylindrical samples with a diameter of 10 mm and a height of 1 mm. The capacity of the sample and the loss tangent were measured with an E7-28 immittance analyzer at 1 kHz frequency at 0.5 V effective voltage. The piezoelectric coefficient d_{33} was determined on polarized cylindrical samples using the APC YE2730A setup by a quasi-static method. The values of the electromechanical coupling coefficient were calculated by the following equation:

$$k_p = \sqrt{\frac{\delta_p}{a_p + b_p \cdot \delta_p}}, \quad (1)$$

where, a_p, b_p are the coefficients determined of Planar Poisson's Ratio, δ_p is the relative resonance gap. The Planar Poisson's ratio value was determined by the frequencies ratio of the third and first (main) overtones of the planar vibration mode on piezoelectric elements in the form of a disk.

3. Results and Discussion

3.1. Investigation of Debinding Process

The TGA curve showed that when heated to 180.5 °C a sharp mass decrease by 86.82% was observed (Figure 6). This is due to the evaporation of two components: ethylene glycol monobutyl ether (EGBE), isopropanol (IPA), and the polymerization of ethylene glycol to polyethylene glycol. The boiling temperatures of evaporating components are much lower at 171 °C and 80.4 °C, respectively.

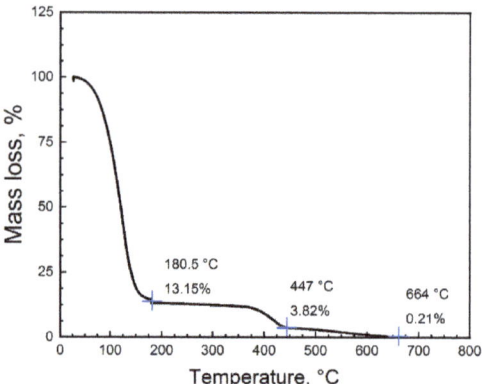

Figure 6. TGA analysis of the binder.

A further mass decrease occurred at a temperature range of 380–450 °C. As a result, the remaining mass of the binder was 0.82% of the initial one. Increasing the temperature leads to a linear decreasing of mass; the binder was almost completely thermally decomposed at 664 °C and the residue was 0.21% of the initial mass. Thus, mass loss of the binder is observed in two stages: the first stage—mass decreases on 86.85%, this stage ends at a temperature of 180.5 °C. The second stage is the temperature range from 180 to 664 °C.

Here, from 180 to 447 °C, no significant mass loss occurs. From 447 to 664 °C, the mass loss is up to 0.21% of the original mass.

The first stage is associated with the transition of ethylene glycol to polyethylene glycol during curing, the second stage is debinding by the burn out of the residue components of the binder [40].

3.2. Binder Jetting Process

For the BJ process, the recoating speed (28 mm/s and 65 mm/s for C-BaTiO$_3$ and F-BaTiO$_3$ respectively) and the frequency of the oscillator (5000 rpm and 4400 rpm for C-BaTiO$_3$ and F-BaTiO$_3$ respectively) were previously optimized to apply a sufficient amount of material to form a smooth thin powder layer. Considering this, the layer thickness for C-BaTiO$_3$ and F-BaTiO$_3$ powder was 100 μm and 35 μm, respectively. The drying time and temperature were also optimized to achieve a uniform layer without cracking and without smearing. The main BJ parameters are shown in Table 1.

Table 1. The main BJ parameters for BaTiO$_3$ lead-free piezoceramic powder.

Process Parameter	C-BaTiO$_3$ Powder	F-BaTiO$_3$ Powder
Recoating speed	28 mm/s	65 mm/s
Frequency of the oscillator	5000 rpm	4400 rpm
Layer thickness	100 μm	35 μm
Drying time	25 s	20 s
Drying temperature	25 °C	33 °C
Roller movement speed	1 mm/s	1 mm/s

Further, the saturation parameter was investigated. Binder saturation is a computed value used to quantify how much binder is dispensed into each unit volume of powder material. Improper saturation of the binder can cause an inhomogeneous layer of powder as well as inaccurate dimensions of printed parts. The theoretical binder saturation (%) was estimated using the following equation:

$$S = \frac{1000 \times V}{(1 - (\frac{PR}{100})) \times X \times Y \times Z}, \quad (2)$$

where V is the volume of binder per drop (pL), PR is the packing rate (%), X and Y are the spacing between binder droplets (μm), and Z is the layer thickness (μm). To obtain the green part with sufficient mechanical strength and surface quality, optimizing the saturation level is critical.

The saturation for C-BaTiO$_3$ powder varied from 40 to 140% with a step of 20%. For F-BaTiO$_3$ powder, the saturation varied from 50 to 200% with a step of 50%. When printing the C-BaTiO$_3$ and F-BaTiO$_3$ samples, no defects were observed on the surface of the powder layer. The powder layer was applied uniformly, the particles did not stick to the roller. After curing of C-BaTiO$_3$ samples printed at 40% saturation, the green model delaminated. At 60% and 80% saturation, the deviation from the computer model size amounted to 0.37 mm along the X and Y-axes and more than 0.1 mm along the Z-axis. For 100% saturation, the C-BaTiO$_3$ samples had clear boundaries and the deviation from the computer model size was about 0.2 mm along the X and Y-axes, and less than 0.05 mm along the Z-axis. At 120% and 140% saturation, the geometry of the green models changed significantly and appeared to be barrel-shaped.

After curing of F-BaTiO$_3$ samples printed at 50% saturation, the green model delaminated since there was not enough binder to bond the layers together. For 100% saturation, the F-BaTiO$_3$ samples had clear boundaries and the deviation from the computer model size was about 0.2 mm along the X and Y-axes, and less than 0.02 mm along the Z-axis. At 150% and 200% saturation, the deviation from the computer model size amounted to

0.36 mm and 0.38 mm along the X and Y-axes and more than 0.21 mm and 0.25 along the Z-axis, respectively.

3.3. Investigation of Sintering Process, Shrinkage, Microstructure, Porosity

For investigation of sintering process $BaTiO_3$ samples, the following samples were selected: C-$BaTiO_3$ samples printed at 60, 80, and 100% saturation; F-$BaTiO_3$ samples printed at 100, 150, and 200% saturation.

To understand the influence of saturation level test-sintering was carried out at 1400 °C with a dwell time of 4 h. Samples of C-$BaTiO_3$ printed with 60 and 80% saturation after test-sintering delaminated due to the weak contact between the layers. Also, F-$BaTiO_3$ samples cracked at 150 and 200% saturation, which is due to the high content of the binder. Seemingly, due to the high content of the binder in samples, during debinding and subsequent sintering, the formation of a large amount of gas occurred, leading to the appearance of cracks. However, the C-$BaTiO_3$ and F-$BaTiO_3$ samples printed at 100% saturation after test-sintering were free from defects. Figure 7 shows an image of F-$BaTiO_3$ samples obtained at different saturations after test-sintering. As a result, samples with 100% saturation for both types of powder were selected for further investigation of the sintering process.

Saturation 200%   Saturation 150%   Saturation 100%

Figure 7. Images of F-$BaTiO_3$ samples printed by BJ with different saturation after sintering. Scale: each sample has a diameter of 15 mm and a height of 10 mm. Heating rate of 10 °C/min to 1400 °C with a dwell time of 4 h.

Subsequently, these samples were subjected to sintering in the temperature range of 1300–1400 °C for 2–6 h. Sintering experiments at temperature 1500 °C led to the melting of $BaTiO_3$ and the destruction of the samples. Initially, the study of the sintering process was carried out for C-$BaTiO_3$ samples at various temperatures of 1300, 1350, and 1400 °C. The best value for the density of the material was achieved at 1400 °C. Considering that the particle size of the C-$BaTiO_3$ powder is close to the particle size of the F-$BaTiO_3$ powder (but different agglomerates sizes), a further investigation of sintering for the F-$BaTiO_3$ samples was carried out at a temperature of 1400 °C with dwell times of 2, 4, and 6 h.

Figure 8 shows graphs of the dependence of sintered samples density on dwell time. The density of the samples increases with an increasing dwell time. The density of F-$BaTiO_3$ samples is higher compared to C-$BaTiO_3$ samples. The density of C-$BaTiO_3$ samples is lower, but the printing speed is higher due to layer thickness difference. During the printing of cylindrical samples from F-$BaTiO_3$ with 15 mm diameter and 10 mm height, the printing time was 7 h, which is 4 h longer than the printing time for similar size samples from C-$BaTiO_3$ powder.

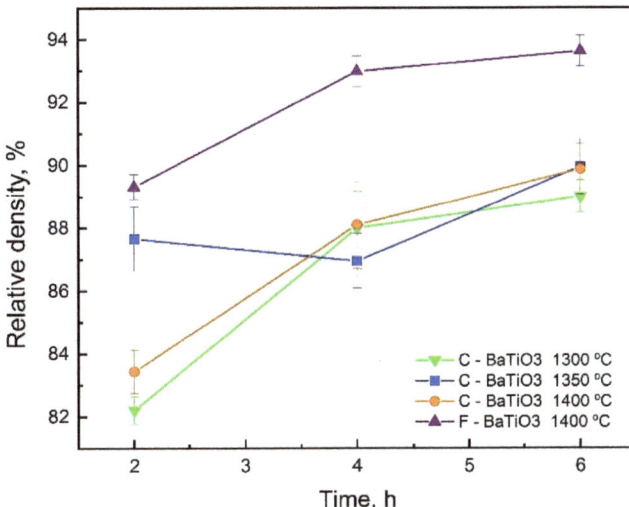

Figure 8. Dependence of the density of samples on temperature and dwell time of sintering.

Increasing the temperature and dwell time of sintering leads to grain enlargement. The graphs in Figure 9 show that the grain size of samples from the unimodal powder is more sensitive to changes in temperature and dwell time compared to samples made from multimodal powder. This feature allows adjusting the functional properties of the material in a wider range.

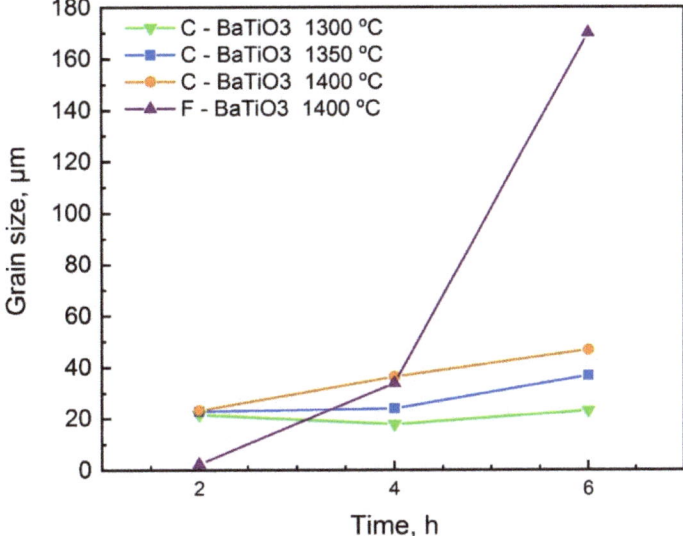

Figure 9. Grain size of BaTiO$_3$ samples dependence on temperature and dwell time of sintering.

As a result of the sintering of samples from C-BaTiO$_3$, the shrinkage along the XY direction was 20–25%. The Z-axis shrinkage varied from 24.1% to 24.4%. The measured linear shrinkage of samples from F-BaTiO$_3$ along the XY direction was 24–27%. The Z-axis shrinkage was 25–26%.

The microstructure of the sintered BaTiO$_3$ samples is shown in Figure 10. The structure is a rounded grain formed as a result of sintering the powder material. Some sintered samples have round-shaped pores, these defects may be associated with binder removal since at this stage there is active gas formation, and perhaps a consequence of the non-optimal sintering mode as well.

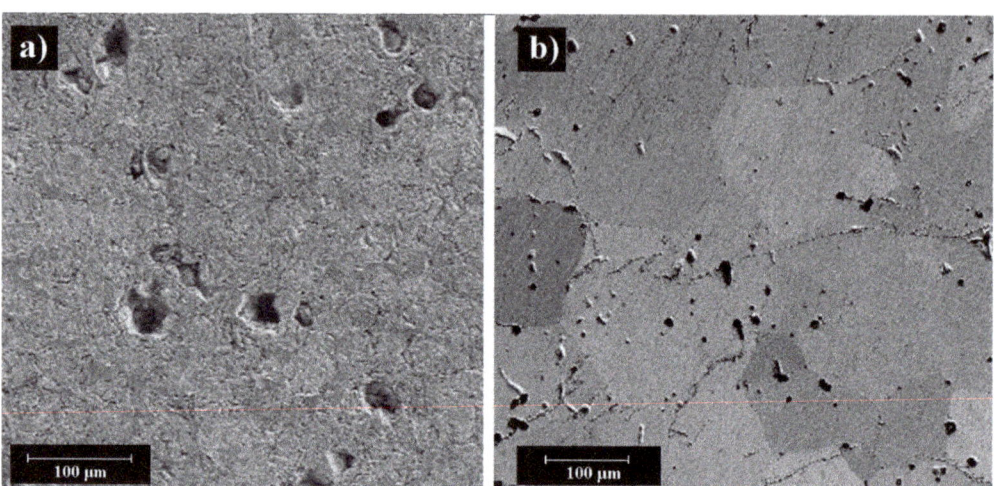

Figure 10. SEM images of sintered samples microstructures at 1400 °C, a dwell time of 6 h from C-BaTiO$_3$ (**a**) and F-BaTiO$_3$ (**b**) powders.

According to EDS measurements, the chemical composition of samples was 59.2% of Ba, 18.8% of Ti, and 22% of O (weight %) which corresponds with BaTiO$_3$ formulation. Figure 11 shows the diffraction patterns of the C-BaTiO$_3$ samples. X-ray diffraction analysis showed that all samples are composed of the tetragonal crystal lattice P4mm of BaTiO$_3$, as evidenced by bifurcated peaks (compare to cubic lattice Pm-3m).

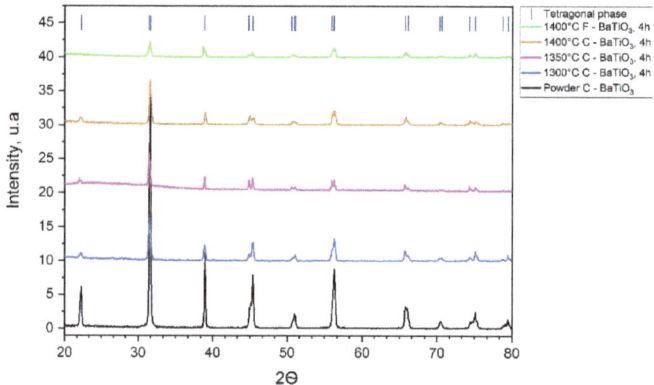

Figure 11. XRD of a sample sintered at different temperatures.

To demonstrate the applicability of the developed modes of the BJ process and thermal post-treatment for manufacturing parts with complex geometries, test samples with lattice structures were made from F-BaTiO$_3$ powder (Figure 12).

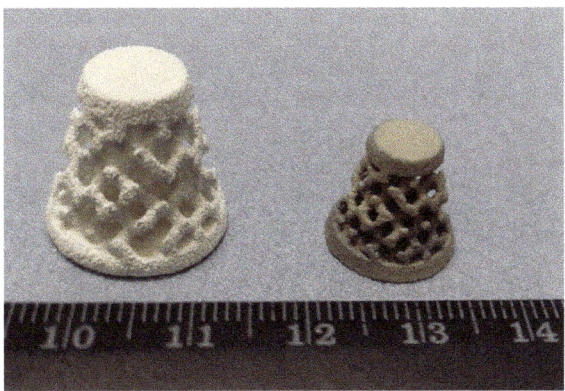

Figure 12. Image of samples with lattice structures printing by BJ from F-BaTiO$_3$ powder before (**left**) and after sintering (**right**).

3.4. Investigation of Functional Properties

The investigation of the functional piezoelectric properties was carried out for C-BaTiO$_3$ samples (a temperature of 1400 °C and a dwell time of 6 h) and F-BaTiO$_3$ (a temperature of 1400 °C and a dwell time of 4 h). These samples were selected considering the highest density and grain size up to 50 microns. This grain size is due to the fact that, for BaTiO$_3$-based piezoceramic, the high functional properties arise with a grain size of 10 to 50 μm [41]. Table 2 shows the test results of the functional properties of sintered samples manufactured from multimodal and unimodal BaTiO$_3$ powders. Samples printed from C-BaTiO$_3$ powder are inferior in dielectric constant, electromechanical coupling coefficient, and piezoelectric coefficient to samples printed from F-BaTiO$_3$. This can be explained by the non-optimal mode of debinding and sintering, the presence of large pores, and as a result, a decrease of the active phase volume of the sample.

Table 2. Piezoelectric properties at 1 kHz of BaTiO$_3$ samples printed by BJ process.

Technology/Powder Type	$\acute{\varepsilon}$	tgδ, %	k_p	d_{33}, pC/N
Binder Jetting/C-BaTiO$_3$	750	5.53	0.15	118
Traditional technology/C-BaTiO$_3$	1872	7.9	0.22	163
Binder Jetting/F-BaTiO$_3$	811	11.59	0.19	183
Traditional technology/F-BaTiO$_3$	2367	1.7	0.36	230

Appreciating the main parameter piezoelectric coefficient d_{33}, it can be noted that using the BJ process allows achieving 72.4% of the piezoelectric coefficient compared to the value obtained by traditional manufacturing technology with multimodal PSD powder and 79.6% of the d_{33} values obtained with unimodal PSD powder. Pressing and sintering were used as the traditional technology, and a solution of polyvinyl alcohol was used as a binder. Sintering was carried out at a temperature of 1350 °C, heating rate 100 °C/h, a dwell time of 3 h.

According to the results of studies published in [32], the functional characteristics of AM piezoceramics depend on the direction of measurement. The functional properties along the Z-axis are about 20% smaller in comparison with the XY direction. In the current study, the properties were measured only along the Z-axis, but the achieved values of piezoelectric coefficient d_{33} = 183 pC/N and dielectric constant $\acute{\varepsilon}$ = 811 exceed the values obtained by the authors [32] parallel (d_{33} = 113 pC/N, $\acute{\varepsilon}$ = 581.6) and perpendicular to the printing orientation (d_{33} = 152.7 pC/N, $\acute{\varepsilon}$ = 698). These differences seem to be related to the raw material and the corresponding difference in technological parameters of BJ and subsequent thermal post-treatment.

The presented results demonstrate that the use of a unimodal PSD powder of lead-free piezoceramics barium titanate allows achieving higher piezoelectric properties, and the use of binder jetting technology allows the creation of objects with complex geometry, which has potential in the manufacture of ultrasonic products used in medicine, aviation, marine industry, sensors for monitoring welded joints, pressure sensors in pipelines, etc.

Future research areas that allow for improving piezoelectric properties include the use of new lead-free piezoelectric materials with increased characteristics (such as KNN, BZT-BCT, etc.), as well as the creation of functional gradient systems and the use of multimaterial 3D printing.

4. Conclusions

The paper presents the results of the additive manufacturing of piezoelectric elements using the binder jetting process. Two powders with different particle size distributions were used as raw materials. Binder jetting with 100% saturation for C-BaTiO$_3$ and for F-BaTiO$_3$ allows printing samples without delamination and cracking. Sintering at 1400 °C with a dwell time of 6 h forms the highest density samples. It was determined that samples from the unimodal powder are more sensitive to increasing grain size during sintering. The measured dielectric and piezoelectric properties of the samples also demonstrated that samples from unimodal powder F-BaTiO$_3$ have higher values. The results of the functional piezoelectric properties obtained by binder jetting with C-BaTiO$_3$ are d_{33} = 118 pC/N, ε = 750, and with F-BaTiO$_3$: d_{33} = 183 pC/N, ε = 811.

The future possibilities of improving functional characteristics of samples manufactured with BJ are increasing speed, optimizing sintering modes, and using new lead-free piezoelectric materials with improved functional characteristics.

Author Contributions: Conceptualization, V.S.; investigation, A.K. and V.S.; methodology, A.S.; project administration, V.S. and A.P.; funding acquisition, A.P.; writing—original draft, A.K. and A.S.; writing—review & editing, A.S. All authors have read and agreed to the published version of the manuscript.

Funding: The research was carried out as part of the work under the State Contract No. Н.4щ.241.09.20.1081 dated 04.06.2020 (ISC 17706413348200001110).

Institutional Review Board Statement: Not applicable.

Informed Consent Statement: Not applicable.

Data Availability Statement: The data presented in this study are available on request from the corresponding author.

Conflicts of Interest: The authors declare no conflict of interest.

References

1. Yin, Q.; Zhu, B.; Zeng, H. *Microstructure, Property and Processing of Functional Ceramics*; Springer: Berlin/Heidelberg, Germany, 2010; ISBN 3642016944.
2. Mostafaei, A.; Elliott, A.M.; Barnes, J.E.; Cramer, C.L.; Nandwana, P.; Chmielus, M. Binder jet 3D printing—Process parameters, materials, properties, and challenges. *Prog. Mater. Sci.* **2020**, 100684. [CrossRef]
3. Eichel, R.-A.; Erünal, E.; Drahus, M.D.; Smyth, D.M.; van Tol, J.; Acker, J.; Kungl, H.; Hoffmann, M.J. Local variations in defect polarization and covalent bonding in ferroelectric Cu^{2+}-doped PZT and KNN functional ceramics at the morphotropic phase boundary. *Phys. Chem. Chem. Phys.* **2009**, *11*, 8698–8705. [CrossRef]
4. Lin, X.; Yuan, F.G. Diagnostic Lamb waves in an integrated piezoelectric sensor/actuator plate: Analytical and experimental studies. *Smart Mater. Struct.* **2001**, *10*, 907. [CrossRef]
5. Chavez, L.A.; Jimenez, F.O.Z.; Wilburn, B.R.; Delfin, L.C.; Kim, H.; Love, N.; Lin, Y. Characterization of thermal energy harvesting using pyroelectric ceramics at elevated temperatures. *Energy Harvest. Syst.* **2018**, *5*, 3–10. [CrossRef]
6. Zhao, X.; Gao, H.; Zhang, G.; Ayhan, B.; Yan, F.; Kwan, C.; Rose, J.L. Active health monitoring of an aircraft wing with embedded piezoelectric sensor/actuator network: I. Defect detection, localization and growth monitoring. *Smart Mater. Struct.* **2007**, *16*, 1208. [CrossRef]
7. Kim, H.; Torres, F.; Villagran, D.; Stewart, C.; Lin, Y.; Tseng, T.B. 3D printing of BaTiO$_3$/PVDF composites with electric in situ poling for pressure sensor applications. *Macromol. Mater. Eng.* **2017**, *302*, 1700229. [CrossRef]

8. Zhang, S.; Yu, F. Piezoelectric materials for high temperature sensors. *J. Am. Ceram. Soc.* **2011**, *94*, 3153–3170. [CrossRef]
9. Chavez, L.A.; Elicerio, V.F.; Regis, J.E.; Kim, H.; Rosales, C.A.G.; Love, N.D.; Lin, Y. Thermal and mechanical energy har-vesting using piezoelectric ceramics. *Mater. Res. Express* **2018**, *6*, 25701. [CrossRef]
10. Yonghong, L.; Zhixin, J.; Jinchun, L. Study on hole machining of non-conducting ceramics by gas-filled electrodischarge and electrochemical compound machining. *J. Mater. Process. Technol.* **1997**, *69*, 198–202. [CrossRef]
11. Cook, R.F.; Freiman, S.W.; Lawn, B.R.; Pohanka, R.C. Fracture of ferroelectric ceramics. *Ferroelectrics* **1983**, *50*, 267–272. [CrossRef]
12. Gao, W.; Zhang, Y.; Ramanujan, D.; Ramani, K.; Chen, Y.; Williams, C.B.; Wang, C.C.L.; Shin, Y.C.; Zhang, S.; Zavattieri, P.D. The status, challenges, and future of additive manufacturing in engineering. *Comput. Des.* **2015**, *69*, 65–89. [CrossRef]
13. Salehi, M.; Gupta, M.; Maleksaeedi, S.; Sharon, N.M.L. *Inkjet Based 3D Additive Manufacturing of Metals*; Materials Research Forum LLC: Millersville, PA, USA, 2018.
14. Xu, Y.; Wu, X.; Guo, X.; Kong, B.; Zhang, M.; Qian, X.; Mi, S.; Sun, W. The boom in 3D-printed sensor technology. *Sensors* **2017**, *17*, 1166. [CrossRef] [PubMed]
15. Zhang, X.; Song, S.; Yao, M.J. Fabrication of embedded piezoelectric sensors and its application in traffic engineering. In Proceedings of the 2017 2nd IEEE International Conference on Intelligent Transportation Engineering (ICITE), Singapore, 1–3 September 2017; pp. 259–265.
16. Bandyopadhyay, A.; Panda, R.K.; McNulty, T.F.; Mohammadi, F.; Danforth, S.C.; Safari, A. Piezoelectric ceramics and compo-sites via rapid prototyping techniques. *Rapid Prototyp. J.* **1998**, *4*, 37–49. [CrossRef]
17. Smay, J.E.; Cesarano, J.; Lewis, J.A. Colloidal inks for directed assembly of 3-D periodic structures. *Langmuir* **2002**, *18*, 5429–5437. [CrossRef]
18. Chen, Y.; Bao, X.; Wong, C.-M.; Cheng, J.; Wu, H.; Song, H.; Ji, X.; Wu, S. PZT ceramics fabricated based on stereolithography for an ultrasound transducer array application. *Ceram. Int.* **2018**, *44*, 22725–22730. [CrossRef]
19. Cui, H.; Hensleigh, R.; Yao, D.; Maurya, D.; Kumar, P.; Kang, M.G.; Priya, S.; Zheng, X.R. Three-dimensional printing of piezoelectric materials with designed anisotropy and directional response. *Nat. Mater.* **2019**, *18*, 234–241. [CrossRef]
20. Lejeune, M.; Chartier, T.; Dossou-Yovo, C.; Noguera, R. Ink-jet printing of ceramic micro-pillar arrays. *J. Eur. Ceram. Soc.* **2009**, *29*, 905–911. [CrossRef]
21. Wan, C.; Bowen, C.R. Multiscale-structuring of polyvinylidene fluoride for energy harvesting: The impact of molecular-, micro-and macro-structure. *J. Mater. Chem. A* **2017**, *5*, 3091–3128. [CrossRef]
22. Kim, H.; Renteria-Marquez, A.; Islam, M.D.; Chavez, L.A.; Garcia Rosales, C.A.; Ahsan, M.A. Fabrication of bulk piezoelectric and dielectric $BaTiO_3$ ceramics using paste extrusion 3D printing technique. *J. Am. Ceram. Soc.* **2018**, *102*, 3685–3694. [CrossRef]
23. Nadkarni, S.S.; Smay, J.E. Concentrated Barium Titanate Colloidal Gels Prepared by Bridging Flocculation for Use in Solid Freeform Fabrication. *J. Am. Ceram. Soc.* **2005**, *89*, 96–103. [CrossRef]
24. Renteria, A.; Diaz, J.A.; He, B.; Renteria-Marquez, I.A.; Chavez, L.A.; Regis, J.E. Particle size influence on material properties of $BaTiO_3$ ceramics fabricated using freeze-form extrusion 3D printing. *Mater. Res. Express* **2019**, *6*, 115211. [CrossRef]
25. Rowlands, W.; Vaidhyanathan, B. Additive manufacturing of barium titanate based ceramic heaters with positive temperature coefficient of resistance (PTCR). *J. Eur. Ceram. Soc.* **2019**, *39*, 3475–3483. [CrossRef]
26. Chen, Z.; Song, X.; Lei, L.; Chen, X.; Fei, C.; Chiu, C.T. 3D printing of piezoelectric element for energy focusing and ultrasonic sensing. *Nano Energy* **2016**, *27*, 78–86. [CrossRef]
27. Cheng, J.; Chen, Y.; Wu, J.-W.; Ji, Z.-R.; Wu, S.-H. 3D Printing of $BaTiO_3$ Piezoelectric Ceramics for a Focused Ultrasonic Array. *Sensors* **2019**, *19*, 4078. [CrossRef]
28. Jang, J.H.; Wang, S.; Pilgrim, S.M.; Schulze, W.A. Preparation and characterization of barium titanate suspensions for stereolithog-raphy. *J. Am. Ceram. Soc.* **2000**, *83*, 1804–1806. [CrossRef]
29. Kim, K.; Zhu, W.; Qu, X.; Aaronson, C.; McCall, W.R.; Chen, S. 3D optical printing of piezoelectric nanoparticle–polymer composite materials. *ACS Nano* **2014**, *8*, 9799–9806. [CrossRef] [PubMed]
30. Sotov, A.; Kantyukov, A.; Popovich, A.; Sufiiarov, V. LCD-SLA 3D printing of $BaTiO_3$ piezoelectric ceramics. *Ceram Int.* **2021**. [CrossRef]
31. Gaytan, S.M.; Cadena, M.A.; Karim, H.; Delfin, D.; Lin, Y.; Espalin, D.; MacDonald, E.; Wicker, R.B. Fabrication of barium titanate by binder jetting additive manufacturing technology. *Ceram Int.* **2015**, *41*, 6610–6619. [CrossRef]
32. Chavez, L.A.; Wilburn, B.R.; Ibave, P.; Delfin, L.C.; Vargas, S.; Diaz, H.; Fulgentes, C.; Renteria, A.; Regis, J.; Liu, Y.; et al. Fabrication and characterization of 3D printing induced orthotropic functional ceramics. *Smart Mater. Struct.* **2019**, *28*, 125007. [CrossRef]
33. Chavez, L.A.; Ibave, P.; Wilburn, B.; Alexander, D.; Stewart, C.; Wicker, R.; Lin, Y. The influence of printing parameters, post-processing, and testing conditions on the properties of binder jetting additive manufactured functional ceramics. *Ceramics* **2020**, *3*, 65–77. [CrossRef]
34. Polley, C.; Distler, T.; Detsch, R.; Lund, H.; Springer, A.; Boccaccini, A.R.; Seitz, H. 3D printing of piezoelectric barium titanate-hydroxyapatite scaffolds with interconnected porosity for bone tissue engineering. *Materials* **2020**, *13*, 1773. [CrossRef] [PubMed]
35. Chen, Z.; Yang, M.; Ji, M.; Kuang, X.; Qi, H.J.; Wang, T. Recyclable thermosetting polymers for digital light processing 3D printing. *Mater. Des.* **2021**, *197*, 109189. [CrossRef]
36. Sufiiarov, V.; Polozov, I.; Kantykov, A.; Khaidorov, A. Binder jetting additive manufacturing of 420 stainless steel: Densification during sintering and effect of heat treatment on microstructure and hardness. *Mater. Today Proc.* **2020**, *30*, 592–595. [CrossRef]

37. Polozov, I.; Sufiiarov, V.; Shamshurin, A. Synthesis of titanium orthorhombic alloy using binder jetting additive manufacturing. *Mater. Lett.* **2019**, *243*, 88–91. [CrossRef]
38. Agapovichev, A.V.; Sotov, A.V.; Kokareva, V.V.; Smelov, V.G.; Kyarimov, R.R. Study of the structure and mechanical characteristics of samples obtained by selective laser melting technology from VT6 alloy metal powder. *Nanosci. Technol. Int. J.* **2017**, *8*, 323–330. [CrossRef]
39. Sufiiarov, V.; Kantyukov, A.; Polozov, I. Reaction sintering of metal-ceramic AlSI-Al_2O_3 composites manufactured by binder jetting additive manufacturing process. In Proceedings of the METAL 2020—29th International Conference on Metallurgy and Materials, Brno, Czech Republic, 20–22 May 2020; pp. 1148–1155. [CrossRef]
40. Do, T.; Kwon, P.; Shin, C.S. Process development toward full-density stainless steel parts with binder jetting printing. *Int. J. Mach. Tools Manuf.* **2017**, *121*, 50–60. [CrossRef]
41. Acosta, M.; Novak, N.; Rojas, V.; Patel, S.; Vaish, R.; Koruza, J.; Rödel, J. $BaTiO_3$-based piezoelectrics: Fundamentals, current status, and perspectives. *Appl. Phys. Rev.* **2017**, *4*, 041305. [CrossRef]

Article

Structure, Mechanical and Magnetic Properties of Selective Laser Melted Fe-Si-B Alloy

Vadim Sufiiarov [1,*], Danil Erutin [1], Artem Kantyukov [1], Evgenii Borisov [1], Anatoly Popovich [1] and Denis Nazarov [2]

[1] Insitiute of Machinery, Materials and Transport, Peter the Great St. Petersburg Polytechnic University, Polytechnicheskaya, 29, 195251 St. Petersburg, Russia; erutin@inbox.ru (D.E.); kantyukov.artem@mail.ru (A.K.); evgenii.borisov@icloud.com (E.B.); director@immet.spbstu.ru (A.P.)

[2] Research Centre "Innovative Technologies of Composite Nanomaterials", St. Petersburg State University, Universitetskaya Nab, 7/9, 199034 St. Petersburg, Russia; dennazar1@yandex.ru

* Correspondence: vadim.spbstu@yandex.ru

Abstract: Original 1CP powder was studied and it was founded that powder material partially consists of the amorphous phase, in which crystallization begins at 450 °C and ends at 575 °C. Selective laser melting parameters were investigated through the track study, and more suitable ones were found: laser power P = 90, 120 W; scanning speed V = 1200 mm/s. Crack-free columnar elements were obtained. The sample obtained with P = 90 W, contains a small amount of amorphous phase. X-ray diffraction of samples shows the presence of α-Fe(Si) and Fe_2B. SEM-image analysis shows the presence of ordered Fe_3Si in both samples. Annealed samples show 40% less microhardness; an annealed sample containing amorphous phase shows higher soft-magnetic properties: 2.5% higher saturation magnetization, 35% higher residual magnetization and 30% higher rectangularity coefficient.

Keywords: selective laser melting; soft-magnetic alloy; FeSiB; magnetic properties; additive manufacturing

Citation: Sufiiarov, V.; Erutin, D.; Kantyukov, A.; Borisov, E.; Popovich, A.; Nazarov, D. Structure, Mechanical and Magnetic Properties of Selective Laser Melted Fe-Si-B Alloy. *Materials* **2022**, *15*, 4121. https://doi.org/10.3390/ma15124121

Academic Editor: Jae Wung Bae

Received: 28 March 2022
Accepted: 7 June 2022
Published: 9 June 2022

Publisher's Note: MDPI stays neutral with regard to jurisdictional claims in published maps and institutional affiliations.

Copyright: © 2022 by the authors. Licensee MDPI, Basel, Switzerland. This article is an open access article distributed under the terms and conditions of the Creative Commons Attribution (CC BY) license (https://creativecommons.org/licenses/by/4.0/).

1. Introduction

Selective laser melting (SLM) technology is an additive manufacturing process by layer-by-layer melting of a 20–60 μm thick powder layer of material using a laser [1,2]. One of the most attractive applications of this technology is the production of bulk amorphous alloys.

Amorphous materials are a type of solid materials in which there is no long-range order in the arrangement of atoms [3]. This state of the material is achieved at high cooling rates from the liquid state due to the fixation of atoms in the positions in which they were in the melt. An amorphous metallic material (metallic glass) does not have a crystalline lattice; therefore, its atomic structure lacks the crystalline defects that cause anisotropy of its properties. Metallic glasses based on ferromagnetic alloys exhibit better soft-magnetic properties than crystalline ferromagnetics: lower coercive force values, higher values of saturation magnetization, higher magnetic permeability and electrical resistivity. Due to the higher level of soft magnetic properties of metallic ferromagnetic glasses, their use as a magnetic core material of an electromagnetic device allows for increasing its efficiency by significantly reducing the magnetic field energy losses for remagnetization and eddy currents [4].

High magnetic properties are achieved with amorphous phases, and materials consisting of nanocrystalline inclusions evenly distributed in the amorphous matrix are also known. Currently, amorphous and nanocrystalline materials produced by additive manufacturing techniques are being investigated by many research groups worldwide [1,4–20]. The problem of obtaining samples with amorphous phase from iron-based soft-magnetic alloys using SLM has been solved with varying degrees of success by selecting process parameters [1,4,11,14,18,19] and by developing and applying specially designed materials and laser beam scanning strategies [4,10,11,19]. Obtaining a material with a minimum degree of crystallinity involves the suppression of crystallization both when cooling the molten metal and when absorbing the heat of the melt by the already solidified part of the product. This task requires researchers to thoroughly understand the impact mechanisms on the resulting material, which makes investigation of the influence of process parameters on the characteristics of the resulting material a priority step towards its solution.

The aim of this work was to investigate the effect of the selective laser melting process parameters (laser power P, hatch distance h, offset m) on macrostructure and microstructure, phase composition, magnetic and mechanical properties of 1CP magnetic alloy samples and to investigate the effect of thermal treatment of samples on their magnetic and mechanical properties.

2. Materials and Methods

The flowability of the powder was determined using ISO 4490 "Determination of flow rate by means of a calibrated funnel (Hall flowmeter)". Apparent density measurements were made by pouring the powder into a funnel from which it flowed into a 25 cm^3 cup. After filling the cup, the funnel was moved away and excess powder was smoothed out with a trowel. Apparent density was determined by weighing the powder in the cup in grams and dividing by 25 cm^3. The skeletal density of the powder was determined in accordance with GOST 22662-77.

The particle size distribution of the powder was determined by laser diffraction on the Analysette 22 NanoTec plus (Fritsch, Germany) with a total measuring range of 0.01–2000 μm. The microstructure of the powder and the obtained samples were studied using a Tescan Mira3 LMU scanning electron microscope (SEM). The etching of the samples was carried out in a 10% nitric acid solution in isopropyl alcohol. The fine structure of the powder was investigated using a Carl Zeiss Libra 200FE transmission electron microscope (TEM) with an energy Ω filter and an operating accelerating voltage of 200 kV. An HAADF detector (STEM mode) and a CCD camera (TEM mode) were used to obtain images. Electron diffraction patterns were measured using an aperture diameter of 1000 nm and a camera length of 450 mm. For the measurement of chemical composition, electron energy loss spectroscopy (EELS) in TEM was used and theoxygen content was measured by infrared absorption and thermal conductivity analysis on LECO TC-500 (LECO Corporation, St. Joseph, MI, USA).

Temperatures of phase transitions were studied using differential scanning calorimeter (DSC) Q2000 (TA Instruments, New Castle, DE, USA) equipped with an automatic sampler, RCS90 cooling system and T-zero baseline alignment technology. Samples were heated in an argon flow to a temperature of 1000 °C at a heating rate of 20 °C/min followed by second heating of the cooled samples to the same temperature. The phase composition was analyzed with a Bruker D8 Advance X-ray diffractometer (XRD) using Cu Kα (l 1/4 1.5418 Å) irradiation.

The magnetic hysteresis loops for the samples were measured by Lake Shore 7410 vibration sample magnetometer (VSM) (Lake Shore Cryotronics, Westerville, OH, USA) at room temperature (22 °C) and under applied different magnetic fields from −18,000 to +18,000 Oe. Magnetic measurements were carried out on 2 sets for each type of sample.

The hardness of the samples was determined using a Buehler Micromet 5103 microhardness tester using the Vickers method at 3 N. To determine the mean value, 5 tests were performed.

Samples were manufactured using an SLM280HL (SLM Solutions GmbH, Lübeck, Germany) selective laser melting system equipped with YLR-Laser (wavelength of 1070 nm and focus size about 80 μm) under a nitrogen atmosphere.

3. Results and Discussion

3.1. Powder Material

The first stage of the research was to investigate the morphology, phase composition, physical, technological and magnetic properties of the initial 1CP powder, consisting of iron and alloying elements: boron, carbon and silicon [21]. The chemical composition of the initial powder is presented in Table 1. The technology of the 1CP powder manufacturing was gas atomization [22].

Table 1. The chemical composition of 1CP powder.

Fe, %	C, %	Si, %	B, %	O, %
Bal.	0.05	2.38	6.42	0.03

The SEM image in Figure 1 shows that the powder material is spherical and rounded particles.

Figure 1. SEM image of 1CP powder.

The results of the particle size distribution of the powder are shown in Table 2. The initial powder particle size is Gaussian distributed with a mean value of 41.8 μm. This is a typical range for use in selective laser melting [18,23].

Table 2. Particle size distribution of 1CP powder.

d_{10}, μm	d_{50}, μm	d_{90}, μm
13.6	41.8	75.3

The results of the investigation of the physical and technological properties of the powder are shown in Table 3. The ability to flow freely through the Hall funnel indicates the possibility of good powder spreading during the formation of thin powder layers in the selective laser melting process. The apparent density is 56.8% of the skeletal density, which indicates an acceptable packing density formed by this powder during the formation of the powder layer.

Table 3. Physical and technological properties of the 1CP powder.

Flow Rate, s/50 g	Apparent Density, g/cm^3	Skeletal Density, g/cm^3
13	4.21	7.41

The X-ray diffraction pattern of the powder sample is shown in Figure 2. The following phases are present in the sample: solid solution α-Fe(Si) and iron boride Fe$_2$B.

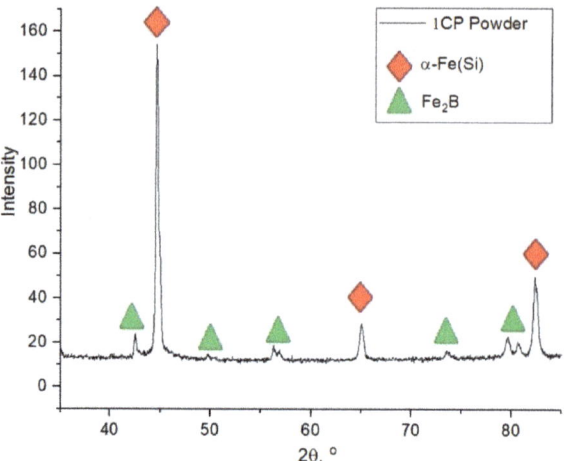

Figure 2. X-ray diffraction pattern of 1CP powder.

Figure 3 shows the DSC results of the powder material presented by two curves: the red curve for the primary heating of the original material and the blue one for the secondary heating of the material (cooled down after primary heating). The primary heating curve shows peaks indicating a phase transformation during heating. This process occurs for 1CP alloy powder in the temperature range from 450 to 575 °C. The absence of the secondary heating peaks shows the accordance of the peaks to the crystallization process. Based on the DSC data it could be concluded that heating above 450 °C would lead to the beginning of crystallization processes of the amorphous phase in case of the presence of the amorphous phase in the samples during heat treatment of samples obtained from this powder. According to this data, it was decided to use annealing heat treatment of samples at 440 °C. This annealing temperature corresponds to the recommended temperature for amorphous ribbons from 1CP [21] and the heat treatment mode used in further study: heating at a rate of 10 °C/min to 440 °C, holding for 30 min, cooling outside the furnace. The halo, which is not clearly visible in the diffraction pattern (Figure 2), is partially visible in the region of 2Θ = 42–47. The absence of an obvious halo can be explained by the small volume content of the amorphous phase in the powder material.

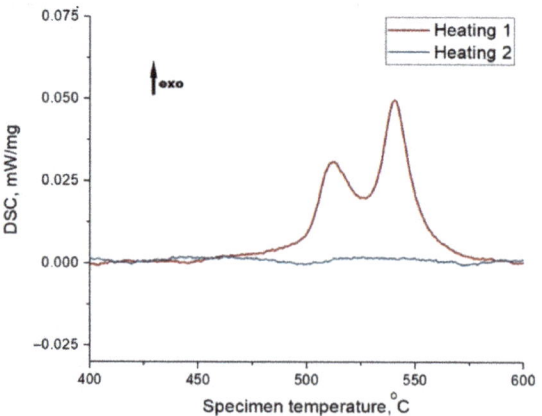

Figure 3. DSC heating curves for the 1CP powder ("exo" means that presented peaks are corresponded to exothermic process).

Figure 4 shows the results of investigation microstructure of 1CP powder by transmission electron microscopy.

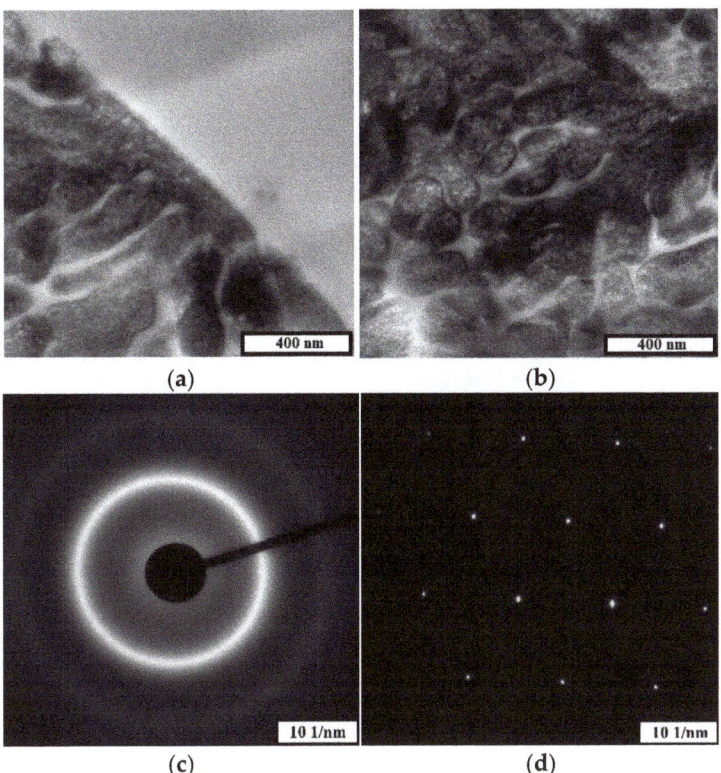

Figure 4. High-resolution TEM images of 1CP metal powder particles (**a**,**b**), electron diffraction pattern of the amorphous phase referring to the light region (**c**), electron diffraction pattern of the crystalline phase referring to the dark region (**d**).

Two phases are present in the studied powder, one of which has a crystalline structure as evidenced by electron diffraction (Figure 4d) and the other has an amorphous structure as evidenced by electron diffraction in Figure 4c. The amorphous phase is present both as separate areas (upper part of Figure 4a) and as areas distributed around the crystalline phase (Figure 4b and lower part of Figure 4a).

The results of the study on the magnetic properties of the powder are shown in Table 4. The hysteresis loop of the powder is shown in Figure 5. 1CP can be considered as a soft magnet with a relatively high coercive force and a low residual magnetization, but a huge saturation magnetization.

Table 4. Magnetic properties of the 1CP powder.

Coercivity, Oe	Saturation Magnetization emu/g	Residual Magnetization emu/g	Rectangularity Factor
33 ± 1	180 ± 3	1.55 ± 0.2	0.008

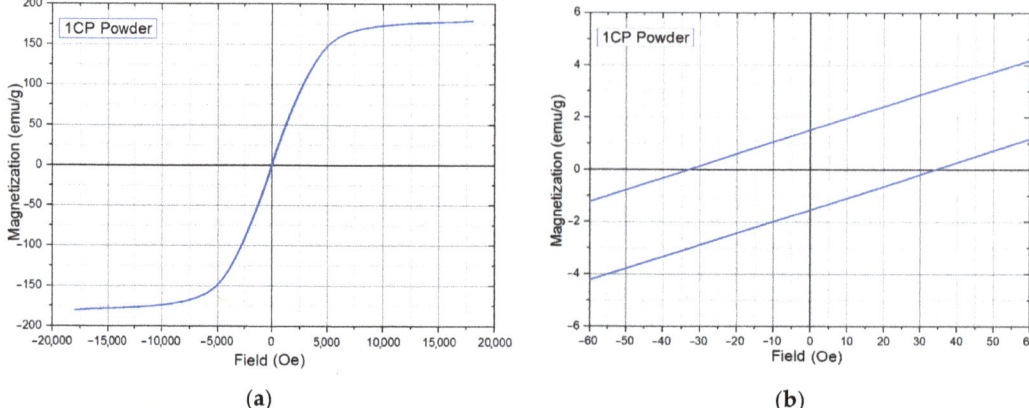

Figure 5. Magnetic properties measurement results of the 1CP powder: (**a**)—general view of hysteresis loop; (**b**)—enlarged area for coercivity estimation.

3.2. Single Track Study

In order to determine the range of applicability parameters for the selective laser melting process, a series of single tracks were melted on a 1CP substrate using different values of laser power P and scanning speed V, which were selected after the preliminary tests have been made with various values of laser power and scanning speed and provided continuous tracks.

The modes used for single-track series are presented in Table 5. The linear energy density is calculated as the ratio of a laser beam power to a scanning speed.

Table 5. Single track modes.

Sample Number	P, W	V, mm/s	Linear Energy Density, J/m
1	60	800	75
2	60	1000	60
3	60	1200	50
4	90	1200	75
5	120	1200	100

The SEM images of the tracks presented in Figure 6 show that there are transverse cracks repeated at distances greater than or close to 200–300 μm. A similar pattern is observed for all the tracks. Therefore, it was decided not to use values of one pass laser length exceeding 200 μm in the next experiments.

Figure 7 shows SEM images of the structure of the melted tracks in a cross-sectional view. The geometric characteristics of the resulting tracks are shown in Table 6. Track 1 has acceptable geometrical characteristics, but there is a pore in its cross-section and a crack at the border with the substrate. The linear energy density of the mode of this track is 75 J/m, as well as of track 4, which has good geometrical characteristics and has no visible defects, but the scanning speed used in the growth of the first track was too low for the used power, which led to the formation of defects. Tracks 2 and 3 were formed at lower linear energy densities (60 J/m and 50 J/m), which were insufficient to make a track with acceptable deposit height.

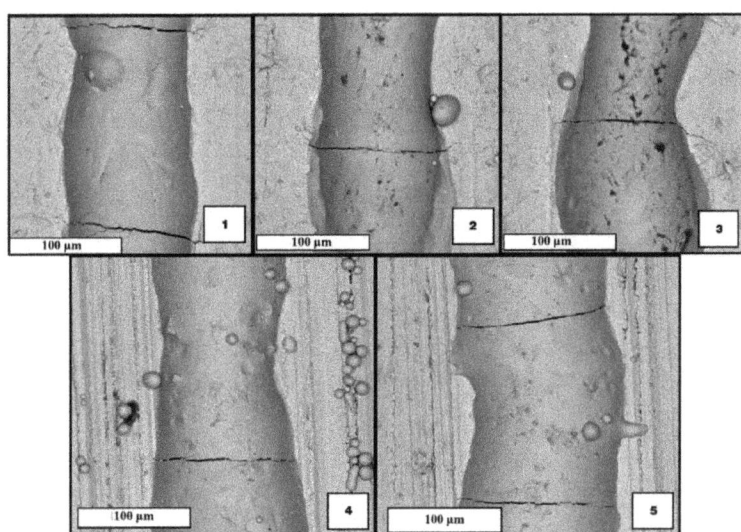

Figure 6. SEM images of top-views single tracks (numbers **1–5** denotes the building mode of Table 5).

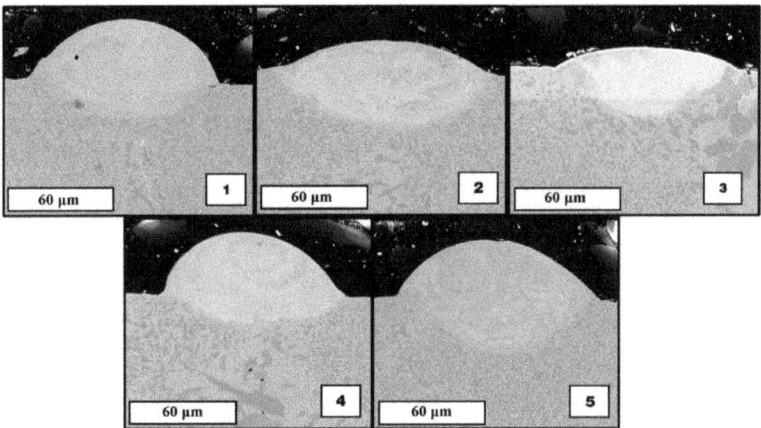

Figure 7. SEM images of cross-sectional views single tracks (numbers **1–5** denotes the building mode of Table 5).

Table 6. Geometric characteristics of the tracks.

Sample	Width, μm	Fusion Depth, μm	Deposit Height, μm
1	108.7	23.1	33.8
2	126.4	33.4	16.8
3	105.6	28.2	12.7
4	105.7	20.6	25.6
5	113.1	31.2	33.1

Tracks 4 and 5 have good geometrical characteristics (sufficient height of the deposited metal and penetration depth) and no visible defects, so modes 4 and 5 are used in the next experiments.

Thus, it was decided to use a melt track length not exceeding 200 μm, with values of $P = 90$ W, $P = 120$ W and $V = 1200$ mm/s.

3.3. Selective Laser Melting of Samples Investigation

As part of the study, eight rectangular samples were manufactured. Samples have been successively made in a nitrogen atmosphere. The plane orthogonal to the height of the cube was divided into cells, each containing two cross-sections of columnar elements, the distance between the centers of which corresponds to the hatch distance parameter h, with the distance between cells corresponding to the offset parameter m. The length of one pass of the laser beam also corresponds to the parameter h. The building scheme is presented in Figure 8.

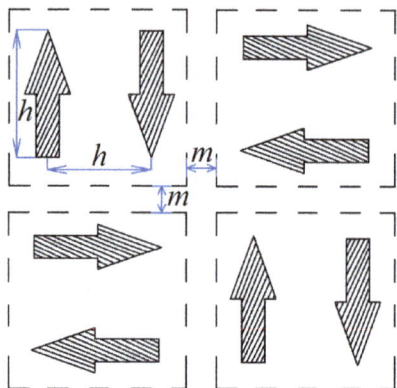

Figure 8. Sample building scheme (arrows in the cells indicates the scanning direction).

The image of manufactured samples is shown in Figure 9. The build modes are presented in Table 7, the scanning speed V and the thickness of the powder layer t were fixed $V = 1200$ mm/s, $t = 50$ µm.

Figure 9. Image of samples manufactured by the SLM process from 1CP alloy powder (numbers 1–8 denotes the building mode of Table 7).

Table 7. Modes of selecting laser melting used for manufacturing samples.

Sample	P, W	h, μm	m, μm
1	90	100	50
2	90	100	0
3	90	100	100
4	90	200	200
5	120	100	50
6	120	100	0
7	120	100	100
8	120	200	200

The samples manufactured at P = 90 W (Figure 10, 1–4) are less dense than those made at P = 120 W (Figure 10, 5–8). Increasing laser power allows the formation of larger structural elements due to the melting of a larger volume of initial powder, which leads to the formation of a denser structure. The increasing value of the offset (Figure 10, 1–4; 5–8) is accompanied by a decrease in the density of samples due to a violation of its structural unity caused by the separation of the columnar elements from each other. The hatch distance parameter h determines the presence of a merger of a pair of columnar elements into a single element: the samples obtained at h = 100 μm (Figure 10, 1–3; 5–7) are characterized by united elements, in contrast to the samples obtained at h = 200 μm (Figure 10, 4; 8).

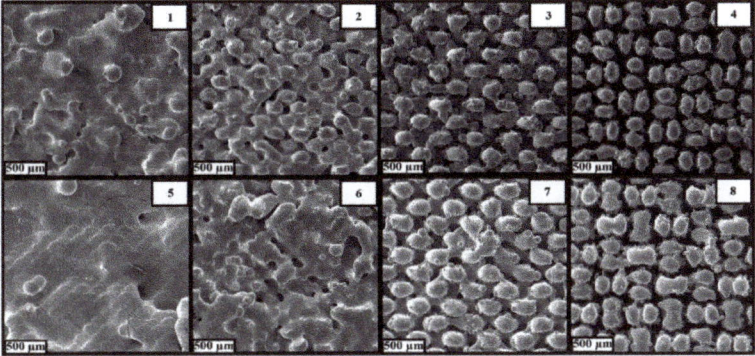

Figure 10. SEM images of samples (top view) produced by selective laser melting with parameters according to Table 6 (numbers **1–8** denotes the building mode of Table 7).

Cross-sectional specimens were prepared for selected columnar elements of samples 4 and 8 (for these samples only the separation of single elements was possible) for examination with a scanning electron microscope. SEM images of the microstructure of the elements are shown in Figures 11 and 12.

The structure of element 8 is characterized by the shape of the layer expressed by the presence of an arc section on the boundary line of each layer. This phenomenon is associated with increased laser power P, the value of which for sample 8 was 120 W. In this case, the change in the shape of the layer is associated with deeper penetration of laser irradiation for a separate section of the layer and uneven distribution of thermal energy over the contact spot of the laser with the metal.

The phase composition was investigated by X-ray diffraction analysis. The X-ray diffraction patterns of the samples are shown in Figure 13.

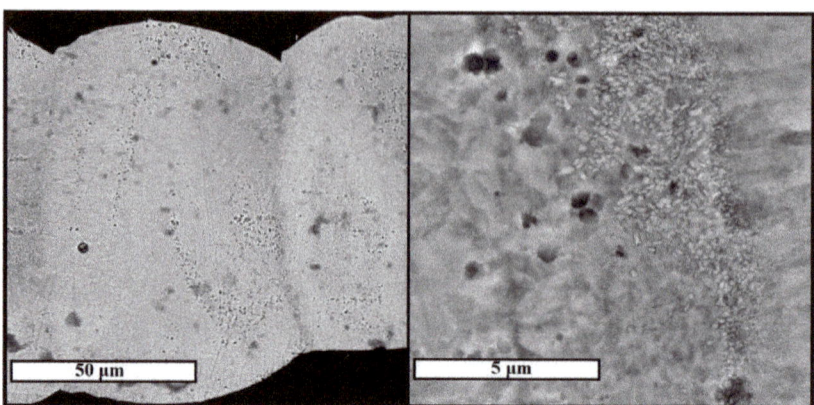

Figure 11. Microstructure of columnar element of sample 4, studied in backscattered electrons mode of SEM.

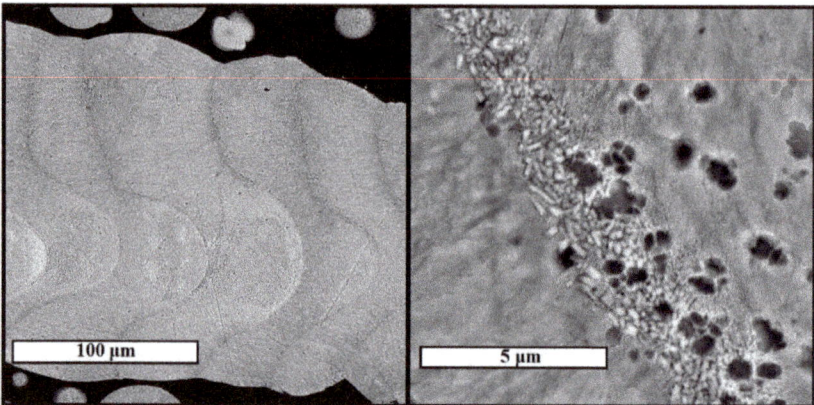

Figure 12. Microstructure of columnar element of sample 8, studied in backscattered electrons mode of SEM.

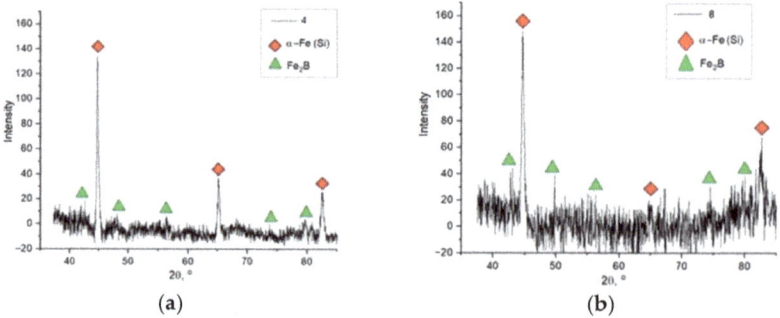

Figure 13. X-ray diffraction patterns of samples 4 (**a**) and 8 (**b**).

Based on X-ray diffraction analysis it was found that the following phases are present in the sample: α-Fe(Si) solid solution and Fe_2B iron boride. The third phase present in the microstructure images of the samples can be identified as an ordered Fe_3Si solid solution. The morphology of the etched cavities is similar to the crystal morphology of this phase [24].

The α-Fe(Si) solid solution has a similar crystallographic structure to the ordered Fe$_3$Si solid solution, due to which the X-ray diffraction analysis may not allow the detection of the reflexes of this phase if the α-Fe(Si) structure prevails [24]. Therefore, researchers [19,23] during the X-ray diffraction analysis of samples of Fe-Si-B alloy obtained by selective laser melting noted the Fe$_3$Si phase together with α-Fe(Si) on the peaks corresponding to α-Fe(Si).

The obtained DSC curves (Figure 14) indicate the almost complete absence of crystallization processes during the heating of the samples. However, the curve of primary heating of sample 4 is characterized by the presence of small peaks, and their absence during secondary heating (which cannot be said for the curve of sample 8), indicating the presence of a small amount of amorphous phase in sample 4.

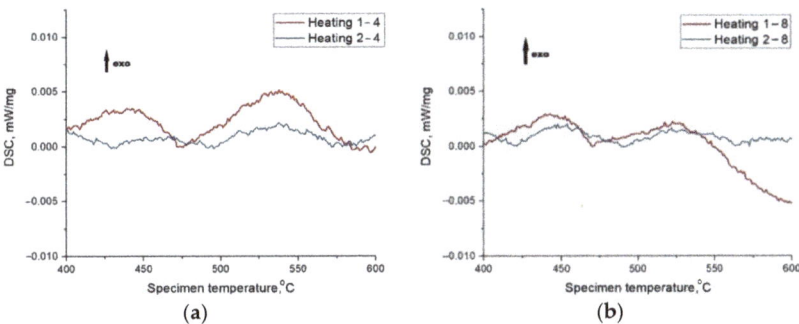

Figure 14. DSC curves of samples 4 (**a**) and 8 (**b**) ("exo" means that presented peaks are corresponded to exothermic process).

Onset crystallization temperatures and enthalpy of the process are presented in Table 8. TEM electron diffraction data presented in Figure 15 proves the presence of the amorphous phase of sample 4.

Table 8. Magnetic properties of the samples.

Sample	Coercivity, Oe	Saturation Magnetization, emu/g	Residual Magnetization, emu/g	Rectangularity Factor
4	49 ± 1	195 ± 3	9.7 ± 0.2	0.050
8	48 ± 1	195 ± 3	10.8 ± 0.2	0.055
4 annealed	48 ± 1	200 ± 3	13.1 ± 0.2	0.065
8 annealed	46 ± 1	195 ± 3	10.2 ± 0.2	0.052

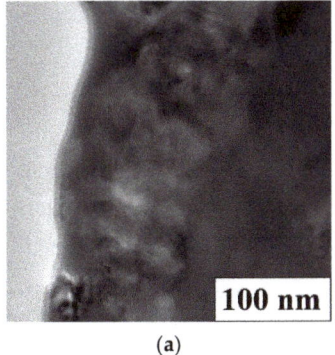
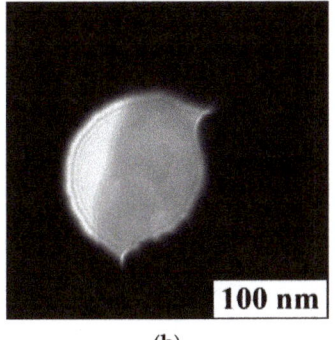
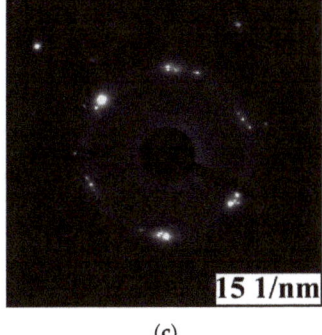

Figure 15. High resolution TEM images of sample 4 (**a**), electron diffraction pattern (**c**) of referring to the region (**b**).

Mechanical and magnetic properties were investigated for the initial and annealed samples. The purpose of annealing was to decrease the level of internal stresses in samples and investigate the effect of it for properties as defined above. The hardness data of the samples (Table 9) indicate that the hardness of sample 4 is slightly higher than that of sample 8. The difference between the mean hardness values is within the standard deviation of the samples ($\sigma_4 = 155$, $\sigma_8 = 92$). Hence, the laser power has no effect on the samples of this material in the investigated power range. The authors [1] investigated samples of a similar composition alloy and obtained hardness values close to those presented in this study. The annealed samples show an approximately 40% reduction values of hardness.

Table 9. Hardness of the samples obtained ($HV_{0.3}$).

Sample	Point 1	Point 2	Point 3	Point 4	Point 5	Mean
4	2226	1998	1986	1841	1854	1981
8	1785	1895	1883	1932	1707	1840
4 annealed	1108	1063	1254	1315	1402	1228
8 annealed	1118	1112	1149	1175	1212	1153

The study of the magnetic properties was carried out for samples 4 and 8. The magnetization curves of the samples are shown in Figure 16. The magnetization curves of the samples after heat treatment are shown in Figure 17. The main parameters of magnetic measurements are summarized in Table 8. The coercivity of the measured samples does not differ significantly from each other. At the same time, there is a difference in the shape of the hysteresis loop (sample 8 achieves a saturation at slightly lower values of field) and the values of residual magnetization. A comparison of the obtained results with the data for amorphous ribbons obtained by melt spinning technology for 1CP alloy [21] shows that the coercivity is much higher and the coefficient of rectangularity is lower for samples made by SLM.

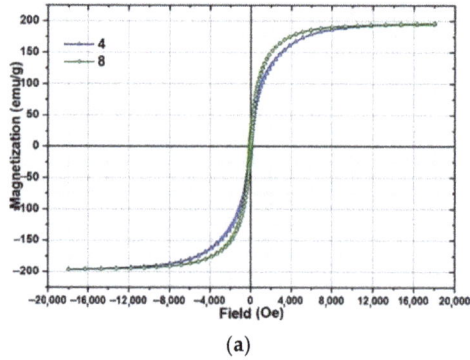
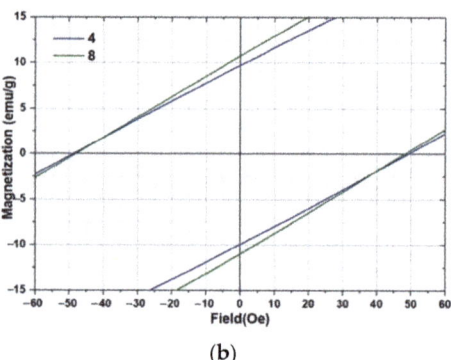

(a)   (b)

Figure 16. Magnetic properties measurements results of samples 4 and 8: (a)—general view of hysteresis loop; (b)—enlarged area for coercivity estimation.

The change of coercivity of annealed samples is within the margin of error. Sample 4 showed higher values of saturation magnetization (2.5% higher), residual magnetization (35% higher) and rectangularity coefficient (30% higher) after annealing. At the same time, sample 8 after heat treatment shows almost the same values of magnetic parameters as before heat treatment. The changing of magnetic properties for sample 4 is possibly related to a relaxation of internal stresses and the presence of a small amount of amorphous phase, magnetic properties changing of which after annealing is stronger than the crystalline phase. Sample 8 has a lower value of internal stresses due to the higher laser power used for its manufacturing and demonstrates no changing magnetic properties after annealing. Therefore, the heat treatment mode recommended for amorphous ribbons of this material [21] should be reevaluated for selective laser melting samples.

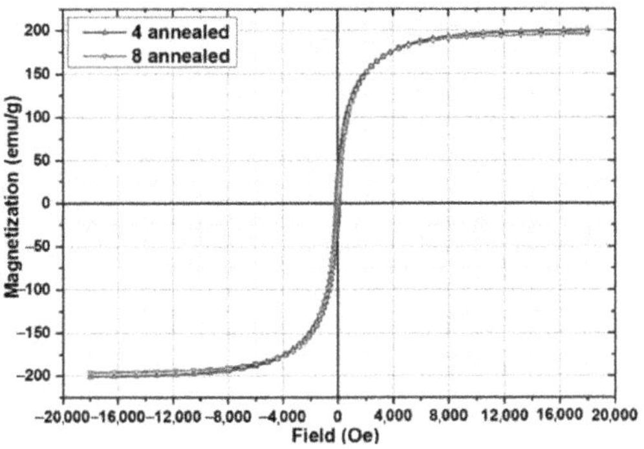

Figure 17. Magnetic properties measurements results of samples 4 and 8 after annealing heat treatment: (a)—general view of hysteresis loop; (b)—enlarged area for coercivity estimation.

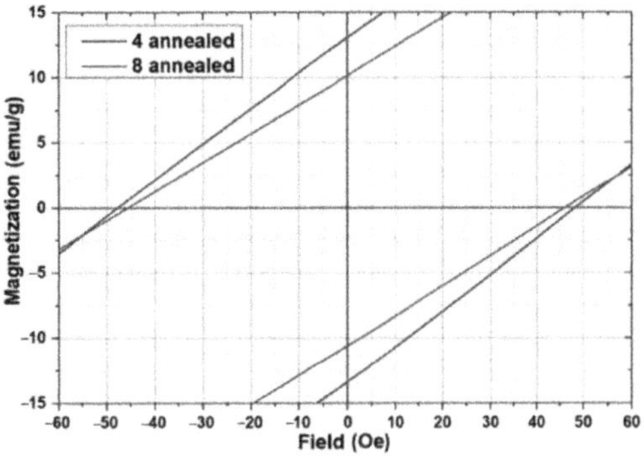

Further research requires the use of scanning strategies with different patterns and multiplicity, substrate heating and cooling experiments, and better optimization of physical and geometric process parameters and reaching more amorphous phase content.

4. Conclusions

The effect of selective laser melting parameters on microstructure and magnetic properties of 1CP alloy was investigated in this work.

1. Increasing the laser power leads to the enlargement of the array elements by melting larger amounts of powder. Increasing the offset parameter of the scanning strategy elements leads to a reduction in the density of the array by moving them farther apart. The hatch distance affected the structural unity of the elements in the pair and, consequently, the density of the array: an increase in this parameter results in a lack of fusion between the elements.
2. Investigation of separate elements of samples 4 and 8 showed no differences in microstructure and phase composition characterized by the presence of solid solution α-Fe(Si), iron boride Fe_2B and, probably, ordered solid solution Fe_3Si. However, sample 8 obtained at $P = 120$ W is characterized by the layer shape expressed by the presence of arc sections on the boundary line, in contrast to sample 4 whose layer boundaries are expressed by smoother curves. Such influence of laser power on interlayers geometry is caused by deeper penetration of laser irradiation and non-uniform distribution of power across a spot of the laser beam.
3. The DSC investigation of samples showed a practically complete absence of crystallization processes in sample 8 in the temperature interval of 400–600 °C in which visible crystallization peaks of initial powder. However, the primary heating curve of sample 4 is characterized by small peaks that indicate the presence of a small amount of amorphous phase in sample 4.
4. The microhardness test results demonstrate no influence of laser beam power on samples manufactured with different laser power. The heat treatment contributes to a decreasing value of microhardness of the samples by about 40%.
5. The study of magnetic properties showed insignificant differences in coercivity and close values of saturation magnetization for samples selective laser melted with different parameters. However, there are differences in hysteresis loop shape and values of residual magnetization: sample 4 has residual magnetization of 9.7 ± 0.2 emu/g, and sample 8 has residual magnetization of 10.8 ± 0.2 emu/g.

Author Contributions: Conceptualization, V.S. and E.B.; methodology, E.B.; validation, A.K.; formal analysis, A.P.; investigation, E.B., D.N. and D.E.; resources, V.S. and A.P.; data curation, A.K. and D.N.; writing—original draft preparation, D.E. and E.B.; writing—review and editing, V.S. and D.N.; visualization, D.E. and A.K.; supervision, V.S. and A.P.; project administration, V.S.; funding acquisition, V.S. All authors have read and agreed to the published version of the manuscript.

Funding: The research was supported by a grant from the Russian Science Foundation № 21-73-10008, https://rscf.ru/project/21-73-10008.

Institutional Review Board Statement: Not applicable.

Informed Consent Statement: Not applicable.

Data Availability Statement: Not applicable.

Acknowledgments: The VSM measurements were conducted using the equipment of the resource centers of the Research Park of the St. Petersburg State University "Innovative Technologies of Composite Nanomaterials".

Conflicts of Interest: The authors declare no conflict of interest.

References

1. Safia, A.; Rima, D.; Nouredine, F. Effect of the Laser Scan Rate on the Microstructure, Magnetic Properties, and Microhardness of Selective Laser-Melted FeSiB. *J. Supercond. Nov. Magn.* **2018**, *31*, 3565–3567.
2. Sufiiarov, V.S.; Popovich, A.A.; Borisov, E.V.; Polozov, I.A. Layer thickness influence on the Inconel 718 alloy microstructure and properties under selective laser melting. *Tsvetnye Met.* **2016**, *2016*, 81–86. [CrossRef]
3. Suzuki, K.; Hujimori, H.; Hashimoto, K. *Amorphous Metals*; Masumoto, T., Ed.; Translated from Japan; Metallurgy: Moscow, Russia, 1987; p. 328.
4. Nam, Y.G.; Koo, B.; Chang, M.S.; Yan, S.; Yu, J.; Park, Y.H.; Jeong, J.W. Selective laser melting vitrification of amorphous soft magnetic alloys with help of double-scanning-induced compositional homogeneity. *Mater. Lett.* **2020**, *261*, 1–4. [CrossRef]
5. Lindroos, T.; Riipinen, T.; Metsä-Kortelainen, S.; Pippuri-Mäkeläinen, J.; Lagerbom, J.; Revuelta, A.; Metsäjoki, J. Soft magnetic alloys for selective laser melting. In Proceedings of the Euro PM2017, Milan, Italy, 1–5 October 2017.
6. Zhang, C.; Ouyang, D.; Pauly, S.; Liu, L. 3D printing of bulk metallic glasses. *Mater. Sci. Eng. R Rep.* **2021**, *145*, 1–43. [CrossRef]
7. Ouyang, D.; Zhang, P.; Zhang, C.; Liu, L. Understanding of crystallization behaviors in laser 3D printing of bulk metallic glasses. *Appl. Mater. Today* **2021**, *23*, 100988. [CrossRef]
8. Pauly, S.; Löber, L.; Petters, R.; Stoica, M.; Scudino, S.; Kühn, U.; Eckert, J. Processing metallic glasses by selective laser melting. *Mater. Today* **2013**, *16*, 37–41. [CrossRef]
9. Nong, X.; Zhou, X.L.; Ren, Y.X. Fabrication and characterization of Fe-based metallic glasses by Selective Laser Melting. *Opt. Laser Technol.* **2019**, *109*, 20–26. [CrossRef]
10. Żrodowski, Ł.; Wysocki, B.; Wróblewski, R.; Kurzydłowski, K.J.; Święszkowski, W. The Novel Scanning Strategy for Fabrication Metallic Glasses by Selective Laser Melting. In Proceedings of the Fraunhofer Direct Digital Manufacturing Conference 2016, Berlin, Germany, 16–17 March 2016.
11. Zou, Y.; Wu, Y.; Li, K.; Tan, C.; Qiu, Z.; Zeng, D. Selective laser melting of crack-free Fe-based bulk metallic glass via chessboard scanning strategy. *Mater. Lett.* **2020**, *272*, 127824. [CrossRef]
12. Ouyang, D.; Wei, X.; Li, N.; Li, Y.; Liu, L. Structural evolutions in 3D-printed Fe-based metallic glass fabricated by selective laser melting. *Addit. Manuf.* **2018**, *23*, 246–252. [CrossRef]
13. Ozden, M.; Morley, N. Laser Additive Manufacturing of Fe-Based Magnetic Amorphous Alloys. *Magnetochemistry* **2021**, *7*, 20. [CrossRef]
14. Jiang, Q.; Zhang, P.; Jie, T.; Zhishui, Y.; Tian, Y.; Ma, S.; Wu, D. Influence of the microstructure on mechanical properties of SLM additive manufacturing Fe-based bulk metallic glasses. *J. Alloys Compd.* **2021**, *894*, 162525. [CrossRef]
15. Liang, S.; Wang, X.; Zhang, W.; Liu, Y.; Wang, W.; Zhang, L. Selective laser melting manufactured porous Fe-based metallic glass matrix composite with remarkable catalytic activity and reusability. *Appl. Mater. Today* **2020**, *19*, 100543. [CrossRef]
16. Shen, Y.; Li, Y.; Chen, C.; Tsai, H.L. 3D printing of large, complex metallic glass structures. *Mater. Des.* **2017**, *117*, 213–222. [CrossRef]
17. Bordeenithikasem, P.; Hofmann, D.; Firdosy, S.; Ury, N.; Vogli, E.; East, D. Controlling microstructure of FeCrMoBC amorphous metal matrix composites via laser directed energy deposition. *J. Alloys Compd.* **2020**, *857*, 157937. [CrossRef]
18. Jung, H.Y.; Choi, S.J.; Konda, G.P.; Mihai, S.; Sergio, S.; Seonghoon, Y.; Uta, K.; Kim, D.H.; Kim, K.B.; Eckertaf, J. Fabrication of Fe-based bulk metallic glass by selective laser melting: A parameter study. *Mater. Des.* **2015**, *86*, 703–708. [CrossRef]
19. Żrodowski, Ł.; Wysocki, B.; Wróblewski, R.; Krawczyńska, A.; Adamczyk-Cieślak, B.; Zdunek, J.; Błyskun, P.; Ferenc, J.; Leonowicz, M.; Święszkowski, W. New approach to amorphization of alloys with low glass forming ability via selective laser melting. *J. Alloys Compd.* **2019**, *771*, 769–776. [CrossRef]

20. Li, Y.; Shen, Y.; Chen, C.; Leu, M.; Tsai, H.L. Building metallic glass structures on crystalline metal substrates by laser-foil-printing additive manufacturing. *J. Mater. Process. Technol.* **2017**, *248*, 249–261. [CrossRef]
21. *TS 14-123-149-2009*; Rapidly Hardened Tape of Soft–Magnetic Amorphous Alloys and Soft–Magnetic Composite Material (Nanocrystalline Alloy), Asha, Russia. 2011.
22. Golod, V.M.; Sufiiarov, V.S. The evolution of structural and chemical heterogeneity during rapid solidification at gas atomization. *IOP Conf. Ser. Mater. Sci. Eng.* **2017**, *192*, 012009. [CrossRef]
23. Gao, S.; Yan, X.; Chang, C.; Aubry, E.; He, P.; Liu, M.; Liao, H.; Fenineche, N. Microstructure and magnetic properties of FeSiBCrC soft magnetic alloy manufactured by selective laser melting. *Mater. Lett.* **2021**, *290*, 129469.
24. Załuska, A.; Matyja, H. Crystallization characteristics of amorphous Fe-Si-B alloys. *J. Mater. Sci.* **1983**, *18*, 2163–2172. [CrossRef]

MDPI
St. Alban-Anlage 66
4052 Basel
Switzerland
Tel. +41 61 683 77 34
Fax +41 61 302 89 18
www.mdpi.com

Materials Editorial Office
E-mail: materials@mdpi.com
www.mdpi.com/journal/materials

www.ingramcontent.com/pod-product-compliance
Lightning Source LLC
LaVergne TN
LVHW070638100526
838202LV00012B/833